MW00709396

Time Series

Time Series
Applications to Finance

Ngai Hang Chan
Chinese University of Hong Kong

A JOHN WILEY & SONS, INC., PUBLICATION

Library of Congress Cataloging-in-Publication Data

Chan, Ngai Hang.
      Time series : applications to finance / Ngai Hang Chan.
          p. cm. — (Wiley series in probability and statistics. Financial engineering section)
      "A Wiley-Interscience publication."
      Includes bibliographical references and index.
      ISBN 0-471-41117-5 (cloth : alk. paper)
          1. Time-series analysis.   2. Econometrics.   3. Risk management.   I. Title.
      II. Series.

      HA30.3 .C47 2002
      332′.01′5195—dc21   2001026955

To Pat and our children, Calvin and Dennis

Contents

Preface   *xi*

1   *Introduction*   *1*
    1.1   *Basic Description*   *1*
    1.2   *Simple Descriptive Techniques*   *5*
       1.2.1   *Trends*   *5*
       1.2.2   *Seasonal Cycles*   *8*
    1.3   *Transformations*   *9*
    1.4   *Example*   *9*
    1.5   *Conclusions*   *13*
    1.6   *Exercises*   *13*

2   *Probability Models*   *15*
    2.1   *Introduction*   *15*
    2.2   *Stochastic Processes*   *15*
    2.3   *Examples*   *17*
    2.4   *Sample Correlation Function*   *18*
    2.5   *Exercises*   *20*

3   *Autoregressive Moving Average Models*   *23*
    3.1   *Introduction*   *23*
    3.2   *Moving Average Models*   *23*
    3.3   *Autoregressive Models*   *25*

		3.3.1	Duality between Causality and Stationarity	26
		3.3.2	Asymptotic Stationarity	28
		3.3.3	Causality Theorem	28
		3.3.4	Covariance Structure of AR Models	29
	3.4	ARMA Models		32
	3.5	ARIMA Models		33
	3.6	Seasonal ARIMA		35
	3.7	Exercises		36

4	Estimation in the Time Domain		39
	4.1	Introduction	39
	4.2	Moment Estimators	39
	4.3	Autoregressive Models	40
	4.4	Moving Average Models	42
	4.5	ARMA Models	43
	4.6	Maximum Likelihood Estimates	44
	4.7	Partial ACF	47
	4.8	Order Selections	49
	4.9	Residual Analysis	53
	4.10	Model Building	53
	4.11	Exercises	54

5	Examples in SPLUS		59
	5.1	Introduction	59
	5.2	Example 1	59
	5.3	Example 2	62
	5.4	Exercises	68

6	Forecasting		69
	6.1	Introduction	69
	6.2	Simple Forecasts	70
	6.3	Box and Jenkins Approach	71
	6.4	Treasury Bill Example	73
	6.5	Recursions	77
	6.6	Exercises	77

7	Spectral Analysis		79
	7.1	Introduction	79
	7.2	Spectral Representation Theorems	79
	7.3	Periodogram	83
	7.4	Smoothing of Periodogram	85
	7.5	Conclusions	89
	7.6	Exercises	89

8   Nonstationarity   93
    8.1   Introduction   93
    8.2   Nonstationarity in Variance   93
    8.3   Nonstationarity in Mean: Random Walk
          with Drift   94
    8.4   Unit Root Test   96
    8.5   Simulations   98
    8.6   Exercises   99

9   Heteroskedasticity   101
    9.1   Introduction   101
    9.2   ARCH   102
    9.3   GARCH   105
    9.4   Estimation and Testing for ARCH   107
    9.5   Example of Foreign Exchange Rates   109
    9.6   Exercises   116

10 Multivariate Time Series   117
    10.1 Introduction   117
    10.2 Estimation of μ and Γ 121
    10.3 Multivariate ARMA Processes   121
          10.3.1 Causality and Invertibility   122
          10.3.2 Identifiability   123
    10.4 Vector AR Models   124
    10.5 Example of Inferences for VAR   127
    10.6 Exercises   135

11 State Space Models   137
    11.1 Introduction   137
    11.2 State Space Representation   137
    11.3 Kalman Recursions   140
    11.4 Stochastic Volatility Models   142
    11.5 Example of Kalman Filtering of Term Structure   144
    11.6 Exercises   150

12 Multivariate GARCH   153
    12.1 Introduction   153
    12.2 General Model   154
          12.2.1 Diagonal Form   155
          12.2.2 Alternative Matrix Form   156
    12.3 Quadratic Form   156
          12.3.1 Single-Factor GARCH(1,1)   156
          12.3.2 Constant-Correlation Model   157

12.4   Example of Foreign Exchange Rates   157
       12.4.1   The Data   158
       12.4.2   Multivariate GARCH in SPLUS   158
       12.4.3   Prediction   166
       12.4.4   Predicting Portfolio Conditional
                Standard Deviations   167
       12.4.5   BEKK Model   168
       12.4.6   Vector-Diagonal Models   169
       12.4.7   ARMA in Conditional Mean   170
12.5   Conclusions   171
12.6   Exercises   171

13 Cointegrations and Common Trends   173
13.1   Introduction   173
13.2   Definitions and Examples   174
13.3   Error Correction Form   177
13.4   Granger's Representation Theorem   179
13.5   Structure of Cointegrated Systems   183
13.6   Statistical Inference for Cointegrated Systems   184
       13.6.1   Canonical Correlations   184
       13.6.2   Inference and Testing   186
13.7   Example of Spot Index and Futures   188
13.8   Conclusions   193
13.9   Exercises   193

References   195

Index   201

Preface

This textbook evolved in conjunction with teaching a course in time series analysis at Carnegie Mellon University and the Chinese University of Hong Kong. For the past several years, I have been involved in developing and teaching the financial time series analysis course for the Masters of Science in Computational Finance Program at Carnegie Mellon University. There are two unique features of this program that differ from those of a traditional statistics curriculum.

First, students in the program have diversified backgrounds. Many of them have worked in the finance world in the past, and some have had extensive trading experiences. On the other hand, a substantial number of these students have already completed their Ph.D. degrees in theoretical disciplines such as pure mathematics or theoretical physics. The common denominator between these two groups of students is that they all want to analyze data the way a statistician does.

Second, the course is designed to be fast paced and concise. Only six weeks of three-hour lectures are devoted to covering the first nine chapters of the text. After completing the course, students are expected to have acquired a working knowledge of modern time series techniques.

Given these features, offering a full-blown theoretical treatment would be neither appropriate nor feasible. On the other hand, offering cookbook-style instruction would never fulfill the intellectual curiosity of these students. They want to attain an intellectual level beyond that required for routine analysis of time series data. Ultimately, these students have to acquire the knack of knowing the conceptual underpinnings of time series modeling in order to get a better understanding of the ever-changing dynamics of the financial world. Consequently, finding an appropriate text which meets these requirements becomes a very challenging task.

As a result, a set of lecture notes that balances theory and applications, particularly within the financial domain, has been developed. The current text is the consequence of several iterations of these lecture notes. In developing the book a number of features have been emphasized.

- The first seven chapters cover the standard topics in statistical time series, but at a much higher and more succinct level. Technical details are left to the references, but important ideas are explained in a conceptual manner. By introducing time series in this way, both students with a strong theoretical background and those with strong practical motivations get excited about the subject early on.

- Many recent developments in nonstandard time series techniques, such as univariate and multivariate GARCH, state space modeling, cointegrations, and common trends, are discussed and illustrated with real finance examples in the last six chapters. Although many of these recent developments have found applications in financial econometrics, they are less well understood among practitioners of finance. It is hoped that the gap between academic development and practical demands can be narrowed through a study of these chapters.

- Throughout the book I have tried to incorporate examples from finance as much as possible. This is done starting in Chapter 1, where an equity-style timing model is used to illustrate the price one may have to pay if the time correlation component is overlooked. The same approach extends to later chapters, where a Kalman filter technique is used to estimate parameters of a fixed-income term-structure model. By giving these examples, the relevance of time series in financial applications can be firmly anchored.

- To the extent possible, almost all of the examples are illustrated through SPLUS programs, with detailed analyses and explanations of the SPLUS commands. Readers will be able to reproduce the analyses by replicating some of the empirical works and testing alternative models so as to facilitate an understanding of the subject. All data and computer codes used in the book are maintained in the Statlib Web site at Carnegie Mellon University and the Web page at the Chinese University of Hong Kong at

  `http://www.sta.cuhk.edu.hk/data1/staff/nhchan/tsbook.html`.

Several versions of these lecture notes have been used in a time series course given at Carnegie Mellon University and the Chinese University of

Hong Kong. I am gratefule for many suggestions, comments, and questions from both students and colleagues at these two institutions. In particular, I am indebted to John Lehoczky for asking me to develop the program, to Jay Kadane for suggesting that I write this book, and to Pantelis Vlachos, who taught part of this course with me during the spring of 2000. Many of the computer programs and examples in Chapters 9 to 13 are contributions by Pantelis. During the writing I have also benefited greatly from consulting activities with Twin Capital Management, in particular Geoffrey Gerber and Pasquale Rocco, for many illuminating ideas from the financial world. I hope this book serves as a modest example of a fruitful interaction between the academic and professional communities. I would also like to thank Ms. Heidi Sestrich for her help in producing the figures in Latex, and to Steve Quigley, Heather Haselkorn, and Rosalyn Farkas, all from Wiley, for their patience and professional assistance in guiding the preparation and production of this book. Financial support from the Research Grant Council of Hong Kong and the National Security Agency throughout this project is gratefully acknowledged. Last, but not least, I would like to thank my wife, Pat Chao, whose contributions to this book went far beyond the call of duty as a part-time proofreader, and to my family, for their understanding and encouragement of my ceaseless transitions between Hong Kong and Pittsburgh during the past five years. Any remaining errors are, of course, my sole responsibility.

Ngai Hang Chan

Shatin, Hong Kong
March 2001

1

Introduction

1.1 BASIC DESCRIPTION

The study of time series is concerned with time correlation structures. It has diverse applications ranging from oceanography to finance. The celebrated CAPM model and the stochastic volatility model are examples of financial models that contain a time series component. When we think of a time series, we usually think of a collection of values $\{X_t : t = 1, \ldots, n\}$ in which the subscript t indicates the time at which the datum X_t is observed. Although intuitively clear, a number of nonstandard features of X_t can be elaborated.

Unequally spaced data (missing values). For example, if the series is about daily returns of a security, values are not available during nontrading days such as holidays.

Continuous-time series. In many physical phenomena, the underlying quantity of interest is governed by a continuously evolving mechanism and the data observed should be modeled by a continuous time series $X(t)$. In finance, we can think of tick-by-tick data as a close approximation to the continuous evolution of the market.

Aggregation. The series observed may represent an accumulation of underlying quantities over a period of time. For example, daily returns can be thought of as the aggregation of tick-by-tick returns within the same day.

Replicated series. The data may represent repeated measurements of the *same quantity* across different subjects. For example, we might monitor the

1

total weekly spending of each of a number of customers of a supermarket chain over time.

Multiple time series. Instead of being a one-dimensional scalar, X_t can be a vector with each component representing an individual time series. For example, the returns of a portfolio that consist of p equities can be expressed as $\mathbf{X}_t = (X_{1t}, \ldots, X_{pt})'$, where each X_{it}, $i = 1, \ldots, p$, represents the returns of each equity in the portfolio. In this case, we will be interested not only in the serial correlation structures within each equity but also the cross-correlation structures among different equities.

Nonlinearity, nonstationarity, and heterogeneity. Many of the time series encountered in practice may behave nonlinearly. Sometimes transformation may help, but we often have to build elaborate models to account for such nonstandard features. For example, the asymmetric behavior of stock returns motivates the study of GARCH models.

Although these features are important, in this book we deal primarily with standard scalar time series. Only after a thorough understanding of the techniques and difficulties involved in analyzing a regularly spaced scalar time series will we be able to tackle some of the nonstandard features.

In classical statistics, we usually assume the X's to be independent. In a time series context, the X's are usually serially correlated, and one of the objectives in time series analysis is to make use of this serial correlation structure to help us build better models. The following example illustrates this point in a confidence interval context.

Example 1.1 *Let X_t be generated by the following model:*

$$X_t = \mu + a_t - \theta a_{t-1}, \quad a_t \sim \mathrm{N}(0,1) \text{ i.i.d.}$$

Clearly, $E(X_t) = \mu$ and $\operatorname{var} X_t = 1 + \theta^2$. Thus,

$$\begin{aligned}
\operatorname{cov}(X_t, X_{t-k}) &= E(X_t - \mu)(X_{t-k} - \mu) \\
&= E(a_t - \theta a_{t-1})(a_{t-k} - \theta a_{t-k-1}) \\
&= \begin{cases} -\theta, & |k| = 1, \\ 1 + \theta^2, & k = 0, \\ 0, & \text{otherwise.} \end{cases}
\end{aligned}$$

Let $\overline{X} = (\sum_{t=1}^{n} X_t)/n$. By means of the formula

$$\operatorname{var}\left(\frac{1}{n}\sum_{t=1}^{n} X_t\right) = \frac{1}{n^2}\sum_{t=1}^{n} \operatorname{var}(X_t) + \frac{2}{n^2}\sum_{t=1}^{n}\sum_{j=1}^{t-1} \operatorname{cov}(X_t, X_j),$$

Table 1.1   Lengths of Confidence Intervals for $n = 50$

θ	$L(\theta)$
-1	$L(-1) = (4 - \frac{2}{50})^{1/2} \cong 2$
-0.5	1.34
0	1
0.5	0.45
1	0.14

it is easily seen that

$$\sigma_{\overline{X}}^2 = \text{var } \overline{X}$$

$$= \frac{1}{n^2} n(1 + \theta^2) - \frac{2}{n^2}(n - 1)\theta$$

$$= \frac{1}{n}\left(1 + \theta^2 - 2\theta + \frac{2\theta}{n}\right)$$

$$= \frac{1}{n}\left[(1 - \theta)^2 + \frac{2\theta}{n}\right].$$

Therefore, $\overline{X} \sim N(\mu, \sigma_{\overline{X}}^2)$. Hence, an approximate 95% confidence interval (CI) for μ is

$$\overline{X} \pm 2\sigma_{\overline{X}}^2 = \overline{X} \pm \frac{2}{\sqrt{n}}\left[(1 - \theta)^2 + \frac{2\theta}{n}\right]^{1/2}.$$

If $\theta = 0$, this CI becomes

$$\overline{X} \pm \frac{2}{\sqrt{n}},$$

coinciding with the independent identically distributed (i.i.d.) case. The difference in the CIs between $\theta = 0$ and $\theta \neq 0$ can be expressed as

$$L(\theta) = \left[(1 - \theta)^2 + \frac{2\theta}{n}\right]^{1/2}.$$

Table 1.1 gives numerical values of the differences for $n = 50$. For example, if $\theta = 1$ and if we were to use a CI of zero for θ, the wrongly constructed CI would be much longer than it is supposed to be. The time correlation structure given by the model helps to produce better inference in this situation. □

Example 1.2 *As a second example, we consider the equity-style timing model discussed in Kao and Shumaker (1999). In this article the authors try to explain the spread between value and growth stocks using several fundamental quantities. Among them, the most interesting variable is the earnings–yield gap reported in Figure 4 of their paper. This variable explains almost 30%*

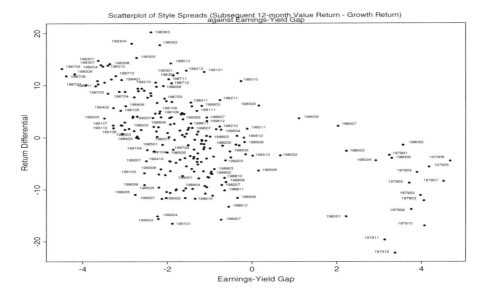

Fig. 1.1   Equity-style timing.

of the variation of the spread between value and growth and suggests that the earnings–yield gap might be a highly informative regressor. Further description of this data set is given in their article. We repeat this particular analysis, but taking into account the time order of the observations. The data between January 79 to June 97 are stored in the file eygap.dat *on the Web page for this book, which can be found at*

http://www.sta.cuhk.edu.hk/data1/staff/nhchan/tsbook.html

For the time being, we restrict our attention to reproducing Figure 4 of Kao and Shumaker (1999). The plot and SPLUS *commands are as follows:*

```
>eyield<-read.table("eygap.dat",header=T)
>plot(eyield[,2],eyield[,3],xlab="Earnings-Yield Gap",
+ ylab="Return Differential")
>title("Scatterplot of Style Spreads (Subsequent
+ 12-month Value Return - Growth Return)
+ against Earnings-Yield Gap, Jan 79- Jun 97",cex=0.6)
>identify(eyield[,2],eyield[,3],eyield[,1],cex=0.5)
```

As illustrated in Figure 1.1, the scattergram can be separated into two clouds, those belonging to the first two years of data and those belonging to subsequent years. When time is taken into account, it seems that finding an $R^2 = 0.3$ depends crucially on the data cloud between 79 and 80 at the lower right-hand corner of Figure 1.1. Accordingly, the finding of such a high *explanatory power from the earnings–yield gap seems to be spurious. This example demonstrates that important information may be missing when the time dimension is not taken properly into account.*   □

1.2   SIMPLE DESCRIPTIVE TECHNIQUES

In general, a time series can be decomposed into a macroscopic component and a microscopic component. The macroscopic component can usually be described through a trend or seasonality, whereas the microscopic component may require more sophisticated methods to describe it. In this section we deal with the macroscopic component through some simple descriptive techniques and defer the study of the microscopic component to later chapters. Consider in general that the time series $\{X_t\}$ is decomposed into a time trend part T_t, a seasonal part S_t, and a microscopic part given by the noise N_t. Formally,

$$X_t = T_t + S_t + N_t$$
$$\cong \mu_t + N_t. \tag{1.1}$$

1.2.1   Trends

Suppose that the seasonal part is absent and we have only a simple time trend structure, so that T_t can be expressed as a parametric function of t, $T_t = \alpha + \beta t$, for example. Then T_t can be identified through several simple devices.

Least squares method. We can use the least squares (LS) procedure to estimate T_t easily [i.e., find α and β such that $\sum (X_t - T_t)^2$ is minimized]. Although this method is convenient, there are several drawbacks.

1. We need to assume a fixed trend for the entire span of the data set, which may not be true in general. In reality, the form of the trend may also be changing over time and we may need an adaptive method to accommodate this change. An immediate example is the daily price of a given stock. For a fixed time span, the prices can be modeled pretty satisfactorily through a linear trend. But everyone knows that the fixed trend will give disastrous predictions in the long run.

2. For the LS method to be effective, we can only deal with a simple restricted form of T_t.

Filtering. In addition to using the LS method, we can filter or smooth the series to estimate the trend, that is, use a smoother or a moving average filter, such as

$$Y_t = \mathrm{Sm}(X_t) = \sum_{r=-q}^{s} a_r X_{t+r}.$$

We can represent the relationship between the output Y_t and the input X_t as

$$X_t \rightarrow \boxed{\text{filter}} \rightarrow \text{Sm}(X_t) = Y_t.$$

The weights $\{a_r\}$ of the filters are usually assumed to be symmetric and normalized (i.e., $a_r = a_{-r}$ and $\sum a_r = 1$). An obvious example is the simple moving average filter given by

$$Y_t = \frac{1}{2q+1} \sum_{r=-q}^{q} X_{t+r}.$$

The length of this filter is determined by the number q. When $q = 1$ we have a simple three-point moving average. The weights do not have to be the same at each point, however. An early example of unequal weights is given by the Spencer 15-point filter, introduced by an English actuary, Spencer, in 1904.

The idea is to use the 15-point filter to approximate the filter that passes through a cubic trend. Specifically, define the weights $\{a_r\}$ as

$$a_r = a_{-r},$$

$$a_r = 0, \qquad |r| > 7,$$

$$(a_0, a_1, \ldots, a_7) = \tfrac{1}{320}(74, \ 67, \ 46, \ 21, \ 3, \ -5, \ -6, \ -3).$$

It can easily be shown that Spencer 15-point filter does not distort a cubic trend; that is, for $T_t = at^3 + bt^2 + ct + d$,

$$\text{Sm}(X_t) = \sum_{r=-7}^{7} a_r T_{t+r} + \sum_{r=-7}^{7} a_r N_{t+r}$$

$$\cong \sum_{r=-7}^{7} a_r T_{t+r}$$

$$= T_t.$$

In general, it can be shown that a linear filter with weights $\{a_r\}$ passes a polynomial of degree k in t, $\sum_{i=0}^{k} c_i t^i$, without distortion if and only if the weights $\{a_r\}$ satisfy two conditions, as described next.

Proposition 1.1 $T_t = \sum_r a_r T_{t+r}$, for all kth-degree polynomials $T_t = c_0 + c_1 t + \cdots + c_k t^k$ if and only if

$$\sum_{r=-s}^{s} a_r = 1,$$

$$\sum_{r=-s}^{s} r^j a_r = 0, \quad \text{for} \ \ j = 1, \ldots, k.$$

The reader is asked to provide a proof of this result in the exercises. Using this result, it is straightforward to verify that Spencer 15-point filter passes a cubic polynomial without distortion. For the time being, let us illustrate the main idea on how a filter works by means of the simple case of a linear trend where $X_t = T_t + N_t$, $T_t = \alpha + \beta t$. Consider applying a $(2q+1)$-point moving average filter (smoother) to X_t :

$$Y_t = \mathrm{Sm}(X_t) = \frac{1}{2q+1} \sum_{r=-q}^{q} X_{t+r}$$

$$= \frac{1}{2q+1} \sum_{r=-q}^{q} [\alpha + \beta(t+r)] + N_{t+r}$$

$$\cong \alpha + \beta t$$

if $\frac{1}{2q+1} \sum_{r=-q}^{q} N_{t+r} \cong 0$. In other words, if we use Y_t to estimate the trend, it does a pretty good job. We use the notation $Y_t = \mathrm{Sm}(X_t) = \hat{T}_t$ and $\mathrm{Res}(X_t) = X_t - \hat{T}_t = X_t - \mathrm{Sm}(X_t) \cong N_t$. In this case we have what is known as a low-pass filter [i.e., a filter that passes through the low-frequency part (the smooth part) and filters out the high-frequency part, N_t]. In contrast, we can construct a high-pass filter that filters out the trend. One drawback of a low-pass filter is that we can only use the middle section of the data. If end-points are needed, we have to modify the filter accordingly. For example, consider the filter

$$\mathrm{Sm}(X_t) = \sum_{j=0}^{\infty} \alpha(1-\alpha)^j X_{t-j},$$

where $0 < \alpha < 1$. Known as the *exponential smoothing technique*, this plays a crucial role in many empirical studies. Experience suggests that α is chosen between 0.1 and 0.3. Finding the best filter for a specific trend was once an important topic in time series. Tables of weights were constructed for different kinds of lower-order trends. Further discussion of this point can be found in Kendall and Ord (1990).

Differencing. The preceding methods aim at estimating the trend by a smoother $\hat{T}_t$. In many practical applications, the trend may be known in advance, so it is of less importance to estimate it. Instead, we might be interested in removing its effect and concentrate on analyzing the microscopic component. In this case it will be more desirable to eliminate or annihilate the effect of a trend. We can do this by looking at the residuals $\mathrm{Res}(X_t) = X_t - \mathrm{Sm}(X_t)$. A more convenient method, however, will be to eliminate the trend from the series directly. The simplest method is differencing. Let B be the backshift operator such that $BX_t = X_{t-1}$. Define

$$\Delta X_t = (1 - B)X_t = X_t - X_{t-1},$$

$$\Delta^j X_t = (1 - B)^j X_t, \ j = 1, 2, \ldots .$$

If $X_t = T_t + N_t$, with $T_t = \sum_{j=0}^{p} a_j t^j$, then $\Delta^j X_t = j! a_j + \Delta^j N_t$ and T_t is eliminated. Therefore, differencing is a form of high-pass filter that filters out the low-frequency signal, the trend T_t, and passes through the high-frequency part, N_t. In principle, we can eliminate any polynomial trend by differencing the series enough times. But this method suffers one drawback in practice. Each time we difference the series, we lose one data point. Consequently, it is not advisable to difference the data too often.

Local curve fitting. If the trend turns out to be more complicated, local curve smoothing techniques beyond a simple moving average may be required to obtain good estimates. Some commonly used methods are spline curve fitting and nonparametric regression. Interested readers can find a lucid discussion about spline smoothing in Diggle (1990).

1.2.2   Seasonal Cycles

When the seasonal component S_t is present in equation (1.1), the methods of Section 1.2.1 have to be modified to accommodate this seasonality. Broadly speaking, the seasonal component can be either additive or multiplicative, according to the following formulations:

$$X_t = \begin{cases} T_t + S_t + N_t, & \text{additive case,} \\ T_t S_t N_t, & \text{multiplicative case.} \end{cases}$$

Again, depending on the goal, we can either estimate the seasonal part by some kind of seasonal smoother or eliminate it from the data by a seasonal differencing operation. Assume that the seasonal part has a period of d (i.e., $S_{t+d} = S_t$, $\sum_{j=1}^{d} S_j = 0$).

(A) *Moving average method.* We first estimate the trend part by a moving average filter running over a complete cycle so that the effect of the seasonality is averaged out. Depending on whether d is odd or even, we perform one of the following two steps:

  1. If $d = 2q$, let $\hat{T}_t = \frac{1}{d} \left(\frac{1}{2} X_{t-q} + X_{t-q+1} + \cdots + X_{t+q-1} + \frac{1}{2} X_{t+q} \right)$ for $t = q+1, \ldots, n-q$.
  2. If $d = 2q+1$, let $\hat{T}_t = \frac{1}{d} \sum_{r=-q}^{q} X_{t+r}$ for $t = q+1, \ldots, n-q$.

After estimating T_t, filter it out from the data and estimate the seasonal part from the residual $X_t - \hat{T}_t$. Several methods are available to attain this last step, the most common being the moving average method. Interested readers are referred to Brockwell and Davis (1991) for further discussions and illustrations of this method. We illustrate this method by means of an example in Section 1.4.

(B) *Seasonal differencing.* On the other hand, we can apply seasonal differencing to eliminate the seasonal effect. Consider the dth differencing

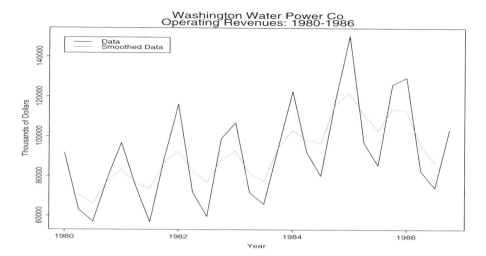

Fig. 1.2   Time series plots.

of the data $X_t - X_{t-d}$. This differencing eliminates the effect of S_t in equation (1.1). Again, we have to be cautious about differencing the data seasonably since we will lose data points.

1.3   TRANSFORMATIONS

If the data exhibit an increase in variance over time, we may need to transform the data before analyzing them. The Box–Cox transformations can be applied here. Experience suggests, however, that log is the most commonly found transformation. Other types of transformations are more problematic, which can lead to serious difficulties in terms of interpretations and forecasting.

1.4   EXAMPLE

In this section we illustrate the idea of using descriptive techniques to analyze a time series. Figure 1.2 shows a time series plot of the quarterly operating revenues of Washington Water Power Company, 1980–1986, an electric and natural gas utility serving eastern Washington and northern Idaho. We start by plotting the data. Several conclusions can be drawn by inspecting the plot.

- As can be seen, there is a slight increasing trend. This appears to drop around 1985–1986.

- There is an annual (12-month) cycle that is pretty clear. Revenues are almost always lowest in the third quarter (July–September) and highest

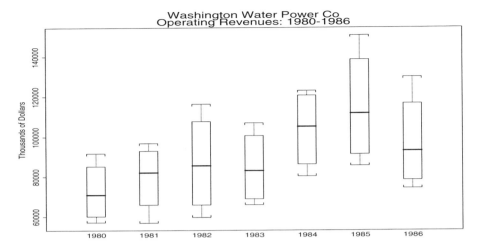

Fig. 1.3   Annual box plots.

in the first quarter (January–March). Perhaps in this part of the country there is not much demand (and hence not much revenue) for electrical power in the summer (for air conditioning, say), but winters are cold and there is a lot of demand (and revenue) for natural gas and electric heat at that time.

- Figure 1.3 shows box plots for each year's operating revenues. The medians seem to rise from year to year and then fall back after the third year. The interquartile range (IQR) gets larger as the median grows and gets smaller as the median falls back; the range does the same. Most of the box plots are symmetric or very slightly positively skewed. There are no outliers.

- In Figure 1.3 we can draw a smooth curve connecting the medians of each year's quarterly operating revenues. We have already described the longer cycle about the medians; this pattern repeats once over the seven-year period graphed. This longer-term cycle is quite difficult to see in the original time series plot.

Assume that the data set has been stored in the file named washpower.dat. The SPLUS program that generates this analysis is listed as follows. Readers are encouraged to work through these commands to get acquainted with the SPLUS program. Further explanations of these commands can be found in the books of Krause and Olson (1997) and Venables and Ripley (1999).

```
>wash_rts(scan(''washpower.dat''),start=1980,freq=4)
>wash.ma_filter(wash,c(1/3,1/3,1/3))
>leg.names_c('Data','Smoothed Data')
```

EXAMPLE 11

```
>ts.plot(wash,wash.ma,lty=c(1,2),
+ main='Washington Water Power Co
Continue string: Operating Revenues: 1980-1986',
+ ylab='Thousands of Dollars',xlab='Year')
>legend(locator(1),leg.names,lty=c(1,2))
>wash.mat_matrix(wash,nrow=4)
>boxplot(as.data.frame(wash.mat),names=as.character(seq(1980,
+ 1986)), boxcol=-1,medcol=1,main='Washington Water Power Co
Continue string: Operating Revenues: 1980-1986',
+ ylab='Thousands of Dollars')
```

To assess the seasonality, we perform the following steps in the moving average method.

1. Estimate the trend through one complete cycle of the series with $n = 28, d = 4$, and $q = 2$ to form $X_t - \hat{T}_t : t = 3, \ldots, 26$. The $\hat{T}_t$ is denoted by `washsea.ma` in the program.

2. Compute the averages of the deviations $\{X_t - \hat{T}_t\}$ over the entire span of the data. Then estimate the seasonal part $\hat{S}_i : i = 1, \ldots, 4$ by computing the demeaned values of these averages. Finally, for $i = 1, \ldots, 4$ let $\hat{S}_{i+4j} = \hat{S}_i : j = 1, \ldots, 6$. The estimated seasonal component $\hat{S}_i$ is denoted by `wash.sea` in the program, and the deseasonalized part of the data $X_t - \hat{S}_t$ is denoted by `wash.nosea`.

3. The third step involves reestimating the trend from the deseasonalized data `wash.nosea`. This is accomplished by applying a filter or any convenient method to reestimate the trend by $\hat{T}_t$, which is denoted by `wash.ma2` in the program.

4. Finally, check the residual $X_t - \hat{T}_t - \hat{S}_t$, which is denoted by `wash.res` in the program, to detect further structures. The SPLUS code follows.

```
> washsea.ma_filter(wash,c(1/8,rep(1/4,3),1/8))
> wash.sea_c(0,0,0,0)
> for(i in 1:2){
+     for(j in 1:6) {
+         wash.sea[i]_wash.sea[i]+
+         (wash[i+4*j][[1]]-washsea.ma[i+4*j][[1]])
+     }
+ }
> for(i in 3:4){
+     for (j in 1:6){
+         wash.sea[i]_wash.sea[i]+
+         (wash[i+4*(j-1)][[1]]-washsea.ma[i+4*(j-1)][[1]])
+     }
+ }
```

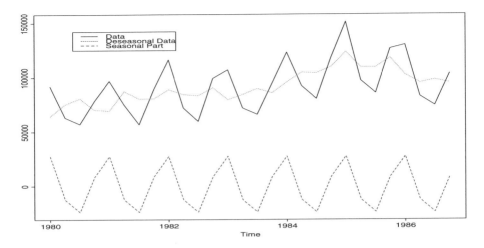

Fig. 1.4 Moving average method of seasonal decomposition.

```
> wash.sea_(wash.sea-mean(wash.sea))/6
> wash.sea1_rep(wash.sea,7)
> wash.nosea_wash-wash.sea
> wash.ma2_filter(wash.nosea,c(1/8,rep(1/4,3),1/8))
> wash.res_wash-wash.ma2-wash.sea
> write(wash.sea1, file='out.dat')
> wash.seatime_rts(scan('out.dat'),start=1980,freq=4)
% This step converts a non-time series object into a time
% series object.
> ts.plot(wash,wash.nosea,wash.seatime)
```

Figure 1.4 gives the time series plot, which contains the data, the deseasonalized data, and the seasonal part. If needed, we can also plot the residual `wash.res` to detect further structures. But it is pretty clear that most of the structures in this example have been identified.

Note that SPLUS also has its own seasonal decomposition function `stl`. Details of this can be found with the `help` command. To execute it, use

```
> wash.stl_stl(wash,'periodic')
> dwash_diff(wash,4)
> ts.plot(wash,wash.stl$sea,wash.stl$rem,dwash)
```

Figure 1.5 gives the plot of the data, the deseasonal part, and the seasonal part. Comparing Figures 1.4 and 1.5 indicates that these two methods accomplish the same task of seasonal adjustments. As a final illustration we can difference the data with four lags to eliminate the seasonal effect. The plot of this differenced series is also drawn in Figure 1.5.

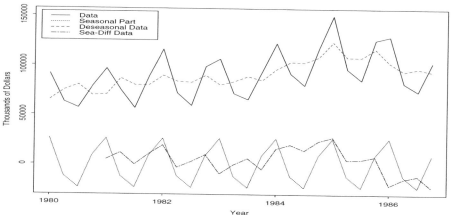

Fig. 1.5 SPLUS `stl` seasonal decomposition.

1.5   CONCLUSIONS

In this chapter we studied several descriptive methods to identify the macroscopic component (trend and seasonality) of a time series. Most of the time, these components can be identified and interpreted easily and there is no reason to fit unnecessarily complicated models to them. From now on we will assume that this preliminary data analysis step has been completed and we focus on analyzing the residual part N_t for microscopic structures. To accomplish this goal, we need to build more sophisticated models.

1.6   EXERCISES

1. (a) Show that a linear filter $\{a_j\}$ passes an arbitrary polynomial of degree k without distortion, that is,

$$m_t = \sum_j a_j m_{t-j},$$

for all kth-degree polynomials $m_t = c_0 + c_1 t + \cdots + c_k t^k$ if and only if

$$\sum_j a_j = 1, \quad \text{and} \quad \sum_j j^r a_j = 0 \ \text{ for } r = 1, ..., k.$$

   (b) Show that the Spencer 15-point moving average filter does not distort a cubic trend.

2. If $m_t = \sum_{k=0}^p c_k t^k, t = 0, \pm 1, ...,$ show that Δm_t is a polynomial of degree $(p-1)$ in t and hence $\Delta^{p+1} m_t = 0$.

3. In SPLUS, get hold of the yearly airline passenger data set by assigning it to an object. You can use the command

```
x_rts(scan('airline.dat'),freq=12,start=1949)
```

The data are now stored in the object x, which forms the time series $\{X_t\}$. This data set consists of monthly totals (in thousands) of international airline passengers from January 1949 to December 1960 [details can be found in Brockwell and Davis (1991)]. It is stored under the file `airline.dat` on the Web page for this book.

(a) Do a time series plot of this data set. Are there any obvious trends?

(b) Is it necessary to transform the data? If a transformation is needed, what would you suggest?

(c) Do a yearly running median for this data set. Sketch the box plots for each year to detect any other trends.

(d) Find a trend estimate by using a moving average filter. Plot this trend.

(e) Estimate the seasonal component S_k, if any.

(f) Consider the deseasonalized data $d_t = X_t - \hat{S}_t, t = 1, ..., n$. Reestimate a trend from $\{d_t\}$ by applying a moving average filter to $\{d_t\}$; call it $\hat{m}_t$, say.

(g) Plot the residuals $r_t = X_t - \hat{m}_t - \hat{S}_t$. Does it look like a white noise sequence? If not, can you make any suggestions?

2

Probability Models

2.1 INTRODUCTION

In the next three chapters, we discuss some theoretical aspects of time series modeling. To gain a better understanding of the microscopic component $\{N_t\}$, basic probability theories of stochastic processes have to be introduced. This is done in the present chapter, and Chapters 3 and 4 deal with commonly used ARIMA models and their basic properties. In Chapter 5, two examples illustrating ideas of these chapters are given in detail with SPLUS commands. Readers who want to become acquainted immediately with series model fitting with SPLUS may want to review some of these examples at this point.

Although one may argue that it is sufficient for a practitioner to analyze a time series without worrying about the technical details, we feel that balanced learning between theory and practice would be much more beneficial. Since time series analysis is a very fast moving field, topics of importance today may become passé in a few years. Thus, it is vital for us to acquire some basic understanding of the theoretical underpinnings of the subject so that when new ideas emerge, we can continue learning on our own.

2.2 STOCHASTIC PROCESSES

Definition 2.1 *A collection of random variables $\{X(t) : t \in \mathcal{R}\}$ is called a* **stochastic process**.

In general, $\{X(t) : 0 \le t < \infty\}$ and $\{X_t : t = 1, 2, \ldots, n\}$ are used to define a continuous-time and a discrete-time stochastic process, respectively. Recall that all the X's are defined on a given probability space $(\Omega, \mathcal{F}, P)$. Thus,

$$X_t = X_t(\omega) : \Omega \to R \quad \text{for a fixed } t.$$

On the other hand, for a given $\omega \in \Omega$, $X_t(\omega)$ can be considered as a function of t and as such, this function is called a *sample function*, a *realization*, or a *sample path* of the stochastic process. For a different ω, it will correspond to a different sample path. The collection of all sample paths is called an *ensemble*. All the time series plots we have seen are based on a single sample path. Accordingly, time series analysis is concerned with finding the probability model that generates the time series observed.

To describe the underlying probability model, we can consider the joint distribution of the process; that is, for any given set of times $(t_1, \ldots, t_n)$, consider the joint distribution of $(X_{t_1}, \ldots, X_{t_n})$, called the *finite-dimensional distribution*.

Definition 2.2 *Let $\mathcal{T}$ be the set of all vectors $\{\boldsymbol{t} = (t_1, \ldots, t_n)' \in T^n : t_1 < \cdots < t_n, \ n = 1, 2, \ldots\}$. Then the* (**finite-dimensional**) **distribution functions** *of the stochastic process $\{X_t, t \in T\}$ are the functions $\{F_{\boldsymbol{t}}(\cdot), \boldsymbol{t} \in \mathcal{T}\}$ defined for $\boldsymbol{t} = (t_1, \ldots, t_n)'$ by*

$$F_{\boldsymbol{t}}(\boldsymbol{x}) = P(X_{t_1} \le x_1, \ldots, X_{t_n} \le x_n), \quad \boldsymbol{x} = (x_1, \ldots, x_n)' \in R^n.$$

Theorem 2.1 *(Kolmogorov's Consistency Theorem) The probability distribution functions $\{F_{\boldsymbol{t}}(\cdot), \boldsymbol{t} \in \mathcal{T}\}$ are the distribution functions of some stochastic process if and only if for any $n \in \{1, 2, \ldots\}, \boldsymbol{t} = (t_1, \ldots, t_n)' \in \mathcal{T}$ and $1 \le i \le n$,*

$$\lim_{x_i \to \infty} F_{\boldsymbol{t}}(\boldsymbol{x}) = F_{\boldsymbol{t}(i)}(\boldsymbol{x}(i)), \tag{2.1}$$

where $\boldsymbol{t}(i)$ and $\boldsymbol{x}(i)$ are the $(n-1)$-component vectors obtained by deleting the ith components of $\boldsymbol{t}$ and $\boldsymbol{x}$, respectively.

This theorem ensures the existence of a stochastic process through specification of the collection of all finite-dimensional distributions. Condition (2.1) ensures a consistency which requires that each finite-dimensional distribution should have marginal distributions that coincide with the *lower* finite-dimensional distribution functions specified.

Definition 2.3 *$\{X_t\}$ is said to be* **strictlystationary** *if for all n, for all $(t_1, \ldots, t_n)$, and for all τ,*

$$(X_{t_1}, \ldots, X_{t_n}) \overset{d}{=} (X_{t_1+\tau}, \ldots, X_{t_n+\tau}),$$

where $\overset{d}{=}$ denotes equality in distribution.

Intuitively, *stationarity* means that the process attains a certain type of statistical equilibrium and the distribution of the process does not change much. It is a very restrictive condition and is often difficult to verify. We next introduce the idea of covariance and a weaker form of stationarity for a stochastic process.

Definition 2.4 *Let* $\{X_t : t \in T\}$ *be a stochastic process such that* $\mathrm{var}(X_t) < \infty$ *for all* $t \in T$. *Then the* **autocovariance function** $\gamma_X(\cdot, \cdot)$ *of* $\{X_t\}$ *is defined by*

$$\gamma_X(r, s) = \mathrm{cov}(X_r, X_s) = E(X_r - EX_r)(X_s - EX_s), \quad r, s \in T.$$

Definition 2.5 $\{X_t\}$ *is said to be* **weakly stationary (second-order stationary, wide-sense stationary)** *if*

(i) $E(X_t) = \mu$ *for all* t.

(ii) $\mathrm{cov}(X_t, X_{t+\tau}) = \gamma(\tau)$ *for all* t *and for all* τ.

Unless otherwise stated, we assume that all moments, $E|X_t|^k$, exist whenever they appear. A couple of consequences can be deduced immediately from these definitions.

1. Take $\tau = 0, \mathrm{cov}(X_t, X_t) = \gamma(0)$ for all t. The means and variances of a stationary process always remain constant.

2. Strict stationarity implies weak stationarity. The converse is not true in general except in the case of a normal distribution.

Definition 2.6 *Let* $\{X_t\}$ *be a stationary process. Then*

(i) $\gamma(\tau) = \mathrm{cov}(X_t, X_{t+\tau})$ *is called the* **autocovariance function**.

(ii) $\rho(\tau) = \gamma(\tau)/\gamma(0)$ *is called the* **autocorrelation function**.

For stationary processes, we expect that both $\gamma(\cdot)$ and $\rho(\cdot)$ taper off to zero fairly rapidly. This is an indication of what is known as the *short-memory behavior* of the series.

2.3 EXAMPLES

1. X_t are i.i.d. random variables. Then

$$\rho(\tau) = \begin{cases} 1, & \tau = 0, \\ 0, & \text{otherwise.} \end{cases}$$

Whenever a time series has this correlation structure, it is known as a *white noise sequence* and the whiteness will become apparent when we study the spectrum of this process.

2. Let Y be a random variable such that var $Y = \sigma^2$. Let $Y_1 = Y_2 = \cdots = Y_t = \cdots = Y$. Then

$$\rho(\tau) = 1 \quad \text{for all } \tau.$$

Hence the process is stationary. However, this process differs substantially from $\{X_t\}$ in example 1. For $\{X_t\}$, knowing its value at one time t has nothing to do with the other values. For $\{Y_t\}$, knowing Y_1 gives the values of all the other Y_t's. Furthermore,

$$\frac{1}{n}(X_1 + \cdots + X_n) \to EX_1 = \mu \quad \text{by the law of large numbers.}$$

But $(Y_1 + \cdots + Y_n)/n = Y$. There is as much randomness in the nth sample average as there is in the first observation for the process $\{Y_t\}$. To prevent situations like this, we introduce the following definition.

Definition 2.7 *If the sample average formed from a sample path of a process converges to the underlying parameter of the process, the process is called* **ergordic**.

For ergordic processes, we do not need to observe separate independent replications of the entire process in order to estimate its mean value or other moments. One sufficiently long sample path would enable us to estimate the underlying moments. In this book, all the time series studied are assumed to be ergodic.

3. Let $X_t = A \cos \theta t + B \sin \theta t$, $A, B \sim (0, \sigma^2)$ i.i.d. Since $EX_t = 0$, it follows that

$$\begin{aligned}
\text{cov}(X_{t+h}, X_t) &= E(X_{t+h} X_t) \\
&= E(A \cos \theta (t+h) + B \sin \theta (t+h))(A \cos \theta t + B \sin \theta t) \\
&= \sigma^2 \cos \theta h.
\end{aligned}$$

Hence the process is stationary.

2.4   SAMPLE CORRELATION FUNCTION

In practice, $\gamma(\tau)$ and $\rho(\tau)$ are unknown and they have to be estimated from the data. This leads to the following definition.

Definition 2.8 *Let $\{X_t\}$ be a given time series and $\overline{X}$ be its sample mean. Then*

(i) $C_k = \sum_{t=1}^{n-k} (X_t - \overline{X})(X_{t+k} - \overline{X})/n$ *is known as the* **sample auto-covariance function** *of X_t.*

(ii) $r_k = C_k/C_0$ *is called the* **sample autocorrelation function** *(ACF).*

The plot r_k versus k is known as a *correlogram*. By definition, $r_0 = 1$. Intuitively, C_k approximates $\gamma(k)$ and r_k approximates $\rho(k)$. Of course, such approximations rely on the ergodicity of the process. Let us inspect the ACF of the following examples. In the exercise, readers are asked to match the sample ACF with the generating time series.

1. For a random series (e.g., Y_t's are i.i.d.), it can be shown that for each fixed k, as the sample size n tends to infinity,

$$r_k \sim AN(0, 1/n);$$

   that is, the random variables $\sqrt{n}\, r_k$ converge in distribution to a standard normal random variable as $n \to \infty$.

   Remark. A statistic T_n is said to be $AN(\mu_n, \sigma_n^2)$ (asymptotic normally distributed with mean μ_n and variance σ_n^2) if $(T_n - \mu_n)/\sigma_n$ converges in distribution to a standard normal random variable as the sample size n tends to infinity. For example, let $X_1, \ldots, X_n$ be i.i.d. random variables with $E(X_1) = \mu$ and $var(X_1) = \sigma^2 > 0$. Let $\overline{X}_n = (\sum_{i=1}^{n} X_i)/n$ and

$$T_n = \frac{\overline{X}_n - \mu}{\sigma/\sqrt{n}}.$$

   Then by the central limit theorem, T_n converges in distribution to a standard normal random variable as $n \to \infty$ [i.e., $\overline{X}_n$ is said to have an $AN(\mu, \sigma^2/n)$]. In this case, $\mu_n = \mu$ and $\sigma_n^2 = \sigma^2/n$.

2. $Y_t \equiv Y, \quad r_k \equiv 1$.

3. A stationary series often exhibits short-term correlation (or short-memory behavior), a large value of ρ_1 followed by a few smaller correlations which subsequently get smaller and smaller.

4. In an alternating series, r_k alternates between positive and negative values. A typical example of such a series is an AR(1) model with a negative coefficient, $Y_t = -\phi Y_{t-1} + Z_t$, where ϕ is a positive parameter and Z_t are i.i.d. random variables.

5. If seasonality exists in the series, it will be reflected in the ACF. In particular, if $X_t = a \cos t\omega$, it can be shown that $r_k \cong \cos k\omega$.

6. In the nonstationary series case, r_k does not taper off for large values of k. This is an indication of nonstationarity and may be caused by many factors.

Notice that the examples above suggest that we can "identify" a time series through inspection of its ACF. Although this sounds promising, it is not a procedure that is always free of error. When we calculate the ACF of any given series with a fixed sample size n, we cannot put too much confidence in the values of r_k for large k's, since fewer pairs of (X_t, X_{t-k}) will be available for computing r_k when k is large. One rule of thumb is not to evaluate r_k for $k > n/3$. Some authors even argue that only r_k's for $k = O(\log n)$ should be computed. In any case, precautions must be taken. Furthermore, if there is a trend in the data, $X_t = T_t + N_t$, then X_t becomes nonstationary (check this with the definition) and the idea of inspecting the ACF of X_t becomes questionable. It is therefore important to detrend the data before interpreting their ACF.

2.5   EXERCISES

1. Given a seasonal series of monthly observations $\{X_t\}$, assume that the seasonal factors $\{S_t\}$ are constants so that $S_t = S_{t-12}$ for all t and assume that $\{Z_t\}$ is a white noise sequence.

   (a) With a global linear trend and additive seasonality, we have $X_t = \alpha + \beta t + S_t + Z_t$. Show that the operator $\Delta_{12} = 1 - B^{12}$ acting on X_t reduces the series to stationarity.

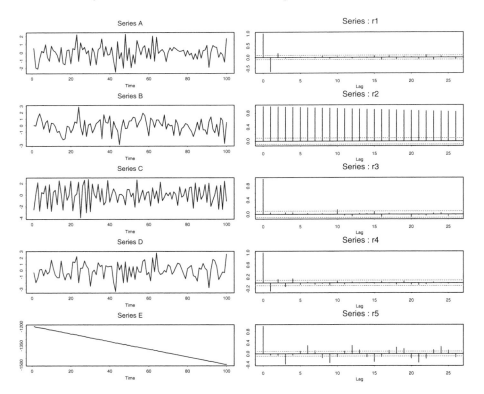

Fig. 2.1   Correlograms for Exercise 3.

(b) With a global linear trend and multiplicative seasonality, we have $X_t = (\alpha + \beta t)S_t + Z_t$. Does the operator Δ_{12} reduce X_t to stationarity? If not, find a differencing operator that does.

2. If $\{X_t = A \cos t\omega : t = 1, \ldots, n\}$ where A is a fixed constant and ω is a constant in $(0, \pi)$, show that $r_k \to \cos k\omega$ as $n \to \infty$. *Hint*: You need to use the double-angle and summation formulas for a trigonometric function.

3. Let $Z_t \sim N(0, 1)$ i.i.d. Match each of the following time series with its corresponding correlogram in Figure 2.1.

   (a) $X_t = Z_t$.

   (b) $X_t = -0.3X_{t-1} + Z_t$.

   (c) $X_t = \sin(\pi/3)\, t + Z_t$.

   (d) $X_t = Z_t - 0.3Z_{t-1}$.

3

Autoregressive Moving Average Models

3.1 INTRODUCTION

Several commonly used probabilistic models for time series analysis are introduced in this chapter. It is assumed that the series being studied have already been detrended by means of the methods introduced in previous chapters. Roughly speaking, there are three kinds of models: the moving average model (MA), the autoregressive model (AR), and the autoregressive moving average model (ARMA). They are used to describe stationary time series. In addition, since certain kinds of nonstationarity can be handled by means of differencing, we also study the class of autoregressive integrated moving average models (ARIMAs).

3.2 MOVING AVERAGE MODELS

Let $\{Z_t\}$ be a sequence of independent identically distributed random variables with mean zero and variance σ^2, denoted by $Z_t \sim$ i.i.d.$(0, \sigma^2)$. If we require $\{Z_t\}$ only to be uncorrelated, not necessarily independent, then $\{Z_t\}$ is sometimes known as a *white noise sequence*, denoted by $Z_t \sim \mathrm{WN}(0, \sigma^2)$. Intuitively, this means that the sequence $\{Z_t\}$ is random with no systematic structures. Throughout this book we use $\{Z_t\}$ to represent a white noise sequence in the loose sense; that is, $\{Z_t\} \sim \mathrm{WN}(0, \sigma^2)$ can mean either $\{Z_t\} \sim$ i.i.d.$(0, \sigma^2)$ or that $\{Z_t\}$ are uncorrelated random variables with mean zero and variance σ^2. By forming a weighted average of Z_t, we end up with a moving average (MA) time series model as follows:

$$Y_t = Z_t + \theta_1 Z_{t-1} + \cdots + \theta_q Z_{t-q}, \quad Z_t \sim \text{WN}(0, \sigma^2). \tag{3.1}$$

This is called a *moving average model of order q*, MA(q). It has many attractive features, including simple mean and autocovariance structures.

Proposition 3.1 *Let $\{Y_t\}$ be the MA(q) model given in (3.1). Then:*

(i) $EY_t = 0$.

(ii) $\text{var } Y_t = (1 + \theta_1^2 + \cdots + \theta_q^2)\sigma^2$.

(iii)

$$\text{cov}(Y_t, Y_{t+k}) = \begin{cases} 0, & \mid k \mid > q, \\ \sigma^2 \displaystyle\sum_{i=0}^{q-|k|} \theta_i \theta_{i+|k|}, & \mid k \mid \leq q. \end{cases}$$

Proof

$$\begin{aligned} \text{cov}(Y_t, Y_{t+k}) &= E(Y_t Y_{t+k}) \\ &= E(Z_t + \cdots + \theta_q Z_{t-q})(Z_{t+k} + \cdots + \theta_q Z_{t+k-q}) \\ &= \sigma^2 \sum_{i=0}^{q-|k|} \theta_i \theta_{i+|k|}, \quad \text{where } \theta_0 \overset{\triangle}{=} 1. \quad\quad \square \end{aligned}$$

Observe that

$$\rho(k) = \begin{cases} \displaystyle\sum_{i=0}^{q-|k|} \theta_i \theta_{i+|k|} \Big/ \sum_{i=0}^{q} \theta_i^2, & \mid k \mid \leq q, \;\; k \neq 0, \\ 1, & k = 0, \\ 0, & \text{otherwise.} \end{cases}$$

Hence, for an MA(q) model, its ACF vanishes after lag q. It is clearly a stationary model. In fact, it can be shown that an MA(q) model is strictly stationary.

Example 3.1 *Consider an MA(1) model $Y_t = Z_t - \theta_1 Z_{t-1}$. Its correlation function satisfies*

$$\rho_Y(k) = \begin{cases} 1, & k = 0, \\ -\theta_1/(1 + \theta_1^2), & \mid k \mid = 1, \\ 0, & \text{otherwise.} \end{cases}$$

Consider another MA(1) model:

$$X_t = Z_t - \frac{1}{\theta_1} Z_{t-1};$$

then

$$\rho_X(k) = \rho_Y(k).$$ □

Both $\{X_t\}$ and $\{Y_t\}$ have the same covariance function. Which one is preferable, $\{X_t\}$ or $\{Y_t\}$? To answer this question, express $\{Z_t\}$ backward in terms of the data. For the data set $\{Y_t\}$, the residual $\{Z_t\}$ can be written as

$$Z_t = Y_t + \theta_1 Z_{t-1} = Y_t + \theta_1(Y_{t-1} + \theta_1 Z_{t-2})$$
$$= Y_t + \theta_1 Y_{t-1} + \theta_1^2 Y_{t-2} + \cdots .$$ (3.2)

For the data set $\{X_t\}$, the residual $\{Z_t\}$ can be written as

$$Z_t = X_t + \frac{1}{\theta_1} Z_{t-1} = \cdots = X_t + \frac{1}{\theta_1} X_{t-1} + \frac{1}{\theta_1^2} X_{t-2} + \cdots .$$ (3.3)

If $|\theta_1| < 1$, equation (3.2) converges and equation (3.3) diverges. When we want to interpret the residuals Z_t, it is more desirable to deal with a convergent expression, and consequently, expression (3.2) is preferable. In this case, the MA(1) model $\{Y_t\}$ is said to be *invertible*.

In general, let $\{Y_t\}$ be an MA(q) model given by $Y_t = \theta(B)Z_t$, where $\theta(B) = 1 + \theta_1 B + \cdots + \theta_q B^q$ with $BZ_t = Z_{t-1}$. The condition for $\{Y_t\}$ to be invertible is given by the following theorem.

Theorem 3.1 *An* MA(q) *model* $\{Y_t\}$ *is invertible if the roots of the equation* $\theta(B) = 0$ *all lie outside the unit circle.*

Proof. The MA(1) case illustrates the idea.   □

Remark. If a constant mean μ is added such that $Y_t = \mu + \theta(B)Z_t$, then $EY_t = \mu$ but the autocovariance function remains unchanged.

3.3 AUTOREGRESSIVE MODELS

Another category of models that is commonly used is the class of autoregressive (AR) models. An AR model has the intuitive appeal that it closely resembles the traditional regression model. When we replace the predictor in the classical regression model by the past (lagged) values of the time series, we have an AR model. It is therefore reasonable to expect that most of the statistical results derived for classical regression can be generalized to the AR case with few modifications. This is indeed the case, and it is for this reason that AR models have become one of the most used linear time series models. Formally, an AR(p) model $\{Y_t\}$ can be written as $\phi(B)Y_t = Z_t$, where $\phi(B) = (1 - \phi_1 B - \cdots - \phi_p B^p)$, $BY_t = Y_{t-1}$, so that

$$Y_t = \phi_1 Y_{t-1} + \cdots + \phi_p Y_{t-p} + Z_t.$$

Formally, we have the following definitions.

Definition 3.1 $\{Y_t\}$ *is said to be an* **AR**(p) *process if:*

   (i) $\{Y_t\}$ is stationarity.

   (ii) $\{Y_t\}$ satisfies $\phi(B)Y_t = Z_t$ for all t.

Definition 3.2 $\{Y_t\}$ *is said to be an* **AR**(p) **process with mean** μ *if* $\{Y_t - \mu\}$ *is an* AR(p) *process.*

3.3.1   Duality between Causality and Stationarity*

There seems to be confusion among different books regarding the notion of stationarity and causality for AR (ARMA in general) models. We clarify this ambiguity in this section.

Main Question: Is it true that an AR(p) always exists?

To answer this question, consider the simple AR(1) case where

$$Y_t = \phi Y_{t-1} + Z_t, \quad Z_t \sim \mathrm{WN}(0, \sigma^2). \tag{3.4}$$

Iterating this equation, $Y_t = Z_t + \phi Z_{t-1} + \cdots + \phi^{k+1}Y_{t-k-1}$. This leads to the following question.

Question 1. Can we find a stationary process that satisfies equation (3.4)?

First, if such a process $\{Y_t\}$ did exist, what would it look like?

- Since $\{Y_t\}$ satisfies (3.4), it must have the following form:

$$Y_t = \sum_{i=0}^{k} \phi^i Z_{t-i} + \phi^{k+1}Y_{t-k-1}.$$

- Assume for the time being that $|\phi| < 1$. Since $\{Y_t\}$ is stationary, $EY_t^2 =$ constant for all t. In particular, denote $||Y_t||^2 = EY_t^2$; we have

$$\left|\left| Y_t - \sum_{j=0}^{k} \phi^j Z_{t-j} \right|\right|^2 = \phi^{2k+2}\left|\left| Y_{t-k-1} \right|\right|^2 \to 0 \text{ as } k \to \infty.$$

Hence, $Y_t = \sum_{j=0}^{\infty} \phi^j Z_{t-j}$ in L^2. For this newly defined process $Y_t = \sum_{j=0}^{\infty} \phi^j Z_{t-j}$, we have the following properties:

---

*Throughout this book, an asterisk indicates a technical section that may be browsed casually without interrupting the flow of ideas.

(i) Y_t satisfies (3.4) for all t.

(ii) $EY_t = 0$,   var $Y_t = \sigma^2/(1 - \phi^2)$.

(iii)

$$\text{cov}\,(Y_t, Y_{t+k}) = \text{cov}\left(\sum_{j=0}^{\infty} \phi^j Z_{t-j}, \sum_{l=0}^{\infty} \phi^l Z_{t+k-l}\right)$$

$$= \sigma^2 \sum_{j=0}^{\infty} \phi^{2j+k} = \sigma^2 \phi^k/(1 - \phi^2).$$

Therefore, the newly defined $\{Y_t\}$ is stationary and the answer to Question 1 is that there exists a stationary AR(1) process $\{Y_t\}$ that satisfies (3.4).

Question 2.   How about the assumption that $|\phi| > 1$?

This assumption is immaterial, since it is not needed once we have established the correct form of $\{Y_t\}$. Although when $|\phi| > 1$, the process $\{Y_t\}$ is no longer convergent, we can rewrite (3.4) as follows. Since $Y_{t+1} = \phi Y_t + Z_{t+1}$, dividing both sides by ϕ, we have

$$Y_t = \frac{1}{\phi} Y_{t+1} - \frac{1}{\phi} Z_{t+1}. \tag{3.5}$$

Replacing t by $t+1$ in (3.5), we arrive at $Y_{t+1} = (Y_{t+2} - Z_{t+2})/\phi$. Substituting this expression into (3.5) and iterating forward on t, we have

$$Y_t = -\frac{1}{\phi} Z_{t+1} + \frac{1}{\phi} Y_{t+1}$$

$$= -\frac{1}{\phi} Z_{t+1} + \frac{1}{\phi}\left(\frac{1}{\phi} Y_{t+2} - \frac{1}{\phi} Z_{t+2}\right)$$

$$= \cdots = -\frac{1}{\phi} Z_{t+1} - \frac{1}{\phi^2} Z_{t+2} - \cdots + \frac{1}{\phi^{k+1}} Y_{t+k+1}.$$

Therefore, $Y_t = -\sum_{j=1}^{\infty} \phi^{-j} Z_{t+j}$, is the stationary solution of (3.4). This process, $\{Z_t\}$, is, however, unnatural since it depends on future values of $\{Y_t\}$, which are unobservable. We have to impose a further condition.

Causal Condition: A useful AR process should depend only on its history [i.e., $\{Z_k : k = -\infty, \dots, t\}$], not on future values. Formally, if there exists a sequence of constants $\{\psi_i\}$ with $\sum_{i=0}^{\infty} |\psi_i| < \infty$ such that $Y_t = \sum_{i=0}^{\infty} \psi_i Z_{t-i}$, the process $\{Y_t\}$ is said to be *causal* (*stationary* in other books).

Question 3. Would the causal condition be too restrictive?

Let $\{Y_t\}$ be the stationary solution of the noncausal AR(1) $Y_t = \phi Y_{t-1} + Z_t$, $|\phi| > 1$. We know that $Y_t = \sum_{j=1}^{\infty} -\phi^{-j} Z_{t+j}$ is a stationary solution to (3.4), albeit noncausal. However, it can be shown that $\{Y_t\}$ also satisfies

$$Y_t = \phi^{-1} Y_{t-1} + \tilde{Z}_t, \quad \tilde{Z}_t \sim (0, \tilde{\sigma}^2)$$

for a newly defined noise $\{\tilde{Z}_t\} \sim$ i.i.d.$(0, \tilde{\sigma}^2)$ (see Exercise 1). Consequently, without loss of generality, we can simply consider causal processes! For the AR(1) case, the causal expression is $Y_t = \sum_{j=0}^{\infty} \phi^j Z_{t-j}$.

3.3.2 Asymptotic Stationarity

There is another subtlety about the AR(1) process. Suppose that the process does not go back to the remote past but starts from an initial value Y_0. Then

$$Y_t = Z_t + \phi Z_{t-1} + \phi^2 Z_{t-2} + \cdots + \phi^{t-1} Z_1 + \phi^t Y_0.$$

If Y_0 is a random variable that is independent of the sequence $\{Z_t\}$ such that $EY_0 \neq 0$, then $EY_t = \phi^t EY_0$. In this case, the process $\{Y_t\}$ is not even stationary. To circumvent this problem, assume that Y_0 is independent of the sequence $\{Z_t\}$ with $EY_0 = 0$. Consider the variance of Y_t:

$$\text{var } Y_t = \sigma^2 \left(1 + \phi^2 + \cdots + \phi^{2(t-1)}\right) + \phi^{2t} \text{ var } Y_0$$

$$= \frac{\sigma^2(1 - \phi^{2t})}{1 - \phi^2} + \phi^{2t} \text{ var } Y_0$$

$$\to \frac{\sigma^2}{(1 - \phi^2)} \quad \text{as } t \to \infty, \quad |\phi| < 1.$$

Even with $EY_0 = 0$, the process $\{Y_t\}$ is nonstationary since its variance is changing over time. It is only stationary when t is large (i.e., it is stationary in an asymptotic sense). With fixed initial values, the AR model is not stationary in the rigorous sense; it is only asymptotically stationary. It is for this reason that when an AR model is simulated, we have to discard the initial chunk of the data so that the effect of Y_0 is negligible.

3.3.3 Causality Theorem

Recall that a process is said to be *causal* if it can be expressed as present and past values of the noise process, $\{Z_t, Z_{t-1}, Z_{t-2}, \dots\}$. Formally, we have the following definition:

Definition 3.3 *A process $\{Y_t\}$ is said to be* **causal** *if there exists a sequence of constants $\{\psi_j\}$'s such that $Y_t = \sum_{j=0}^{\infty} \psi_j Z_{t-j}$ with $\sum_{j=0}^{\infty} |\psi_j| < \infty$.*

For an AR(p) model $\phi(B)Y_t = Z_t$, we write

$$Y_t = \phi^{-1}(B)Z_t = \psi(B)Z_t = \sum_{i=0}^{\infty} \psi_i Z_{t-i}, \tag{3.6}$$

where $\psi_0 = 1$. Under what conditions would this expression be well defined [i.e., would the AR(p) model be causal]? The answer to this question is given by the following theorem.

Theorem 3.2 *An* AR(p) *process is causal if the roots of the characteristic polynomial* $\phi(z) = 1 - \phi z - \cdots - \phi_p z^p$ *are all lying outside the unit circle* [*i.e.,* $\{z : \phi(z) = 0\} \subseteq \{z : |z| > 1\}$].

Proof. Let the roots of $\phi(z)$ be $\zeta_1, \ldots, \zeta_p$. Note that some of them may be equal. By the assumption given, we can arrange their magnitudes in increasing order so that $1 < |\zeta_1| < \cdots < |\zeta_p|$. Write $|\zeta_1| = 1 + \epsilon$ for some $\epsilon > 0$. For z such that $|z| < 1 + \epsilon, \phi(z) \neq 0$. Consequently, there is a power series expansion for $\phi(z)^{-1}$ for $|z| < 1 + \epsilon$; that is,

$$\frac{1}{\phi(z)} = \sum_{i=0}^{\infty} \psi_i z^i, \tag{3.7}$$

is a convergent series for $|z| < 1 + \epsilon$. Now choose $0 < \delta < \epsilon$ and substitute $z = 1 + \delta$ into equation (3.7). Then

$$\frac{1}{\phi(1+\delta)} = \sum_{i=0}^{\infty} \psi_i (1+\delta)^i < \infty.$$

Therefore, there exists a constant $M > 0$ so that for all i,

$$|\psi_i (1+\delta)^i| < M;$$

that is, for all i,

$$|\psi_i| < M(1+\delta)^{-i}.$$

Hence, $\sum_{i=0}^{\infty} |\psi_i| < \infty$. As a result, the process

$$Y_t = \frac{1}{\phi(B)} Z_t = \sum_{i=0}^{\infty} \psi_i Z_{t-i}$$

is well defined and hence causal. □

3.3.4 Covariance Structure of AR Models

Given a causal AR(p) model, we have

$$\gamma(k) = E(Y_t Y_{t+k}) = E\left(\sum_{i=0}^{\infty} \psi_i Z_{t-i}\right)\left(\sum_{l=0}^{\infty} \psi_l Z_{t+k-l}\right)$$

$$= \sigma^2 \sum_{i=0}^{\infty} \psi_i \psi_{k+i} \tag{3.8}$$

Example 3.2 *For an* $\mathrm{AR}(1)$ *model* $Y_t = \phi Y_{t-1} + Z_t$, *we have* $\psi_i = \phi^i$, *so that* $\gamma(k) = \sigma^2 \phi^k / (1 - \phi^2)$ *and* $\rho(k) = \phi^k$.   □

Although we can use equation(3.8) to find the covariance function of a given $\mathrm{AR}(p)$ model, it requires solving ψ's in terms of ϕ's and it is often difficult to find an explicit formula. We can finesse this difficulty by the following observation. Let $\{Y_t\}$ be a given stationary and causal $\mathrm{AR}(p)$ model. Multiply Y_t throughout by Y_{t-k}:

$$Y_t Y_{t-k} = \phi_1 Y_{t-1} Y_{t-k} + \cdots + \phi_p Y_{t-p} Y_{t-k} + Z_t Y_{t-k}.$$

Taking expectations yields

$$\gamma(k) = \phi_1 \gamma(k-1) + \cdots + \phi_p \gamma(k-p).$$

Dividing this equation by $\gamma(0)$, we get

$$\rho(k) = \phi_1 \rho(k-1) + \cdots + \phi_p \rho(k-p) \quad \text{for all } k.$$

We arrive at a set of difference equations, the Yule–Walker equations, whose general solutions are given by

$$\rho(k) = A_1 \pi_1^{-|k|} + \cdots + A_p \pi_p^{-|k|},$$

where $\{\pi_i\}$ are the solutions of the corresponding characteristic equation of the $\mathrm{AR}(p)$ process,

$$1 - \phi_1 z^{p-1} - \cdots - \phi_p z^p = 0.$$

Example 3.3 *Let* $\{Y_t\}$ *be an* $\mathrm{AR}(2)$ *model such that* $Y_t = \phi_1 Y_{t-1} + \phi_2 Y_{t-2} + Z_t$. *The characteristic equation is* $1 - \phi_1 z - \phi_2 z^2 = 0$ *with solutions*

$$\pi_i = \frac{1}{2\phi_2} \left(-\phi_1 \pm \sqrt{\phi_1^2 + 4\phi_2} \right), \quad i = 1, 2.$$

According to Theorem 3.2, the condition $|\pi_i| > 1$, $i = 1, 2$, *guarantees that* $\{Y_t\}$ *be causal. This condition can be shown to be equivalent to the following three inequalities:*

$$\phi_1 + \phi_2 < 1,$$
$$\phi_1 - \phi_2 > -1, \tag{3.9}$$
$$|\phi_2| < 1.$$

To see why this is the case, let the $\mathrm{AR}(2)$ *process be causal so that the characteristic polynomial* $\phi(z) = 1 - \phi_1 z - \phi_2 z^2$ *has roots outside the unit circle. In particular, none of the roots of* $\phi(z) = 0$ *lies between* $[-1, 1]$ *on the real line. Since* $\phi(z)$ *is a polynomial, it is continuous. By the intermediate value theorem,* $\phi(1)$ *and* $\phi(-1)$ *must have the same sign. Otherwise, there* *exists a root of $\phi(z) = 0$ between $[-1, 1]$. Furthermore, $\phi(0)$ has the same sign as $\phi(1)$ and $\phi(-1)$. Now, $\phi(0) = 1 > 0$, so $\phi(1) > 0$ and $\phi(-1) > 0$. Hence,*

$$\phi(1) = 1 - \phi_1 - \phi_2 > 0;$$

that is,

$$\phi_2 + \phi_1 < 1.$$

Also,

$$\phi(-1) = 1 + \phi_1 - \phi_2 > 0;$$

that is,

$$\phi_2 - \phi_1 < 1.$$

To show that $|\phi_2| < 1$, let α and β be the roots of $\phi(z) = 0$ so that $|\alpha| > 1$ and $|\beta| > 1$ due to causality. According to the roots of a quadratic polynomial, we have

$$\alpha\beta = \frac{-1}{\phi_2},$$

$$|\phi_2| = \frac{1}{|\alpha\beta|}$$

$$= \frac{1}{|\alpha||\beta|}$$

$$< 1,$$

showing that causality for an $\mathrm{AR}(2)$ model is equivalent to (3.9).

Given ϕ_1 and ϕ_2, we can solve for π_1 and π_2. Furthermore, if the two roots are real and distinct, we can obtain a general solution of $\rho(k)$ by means of solving a second-order difference equation. Details of this can be found in Brockwell and Davis (1991). The main feature is that the solution $\rho(k)$ consists of a mixture of damped exponentials. In Figure 3.1, plots of the ACF for an $\mathrm{AR}(2)$ model for different values of ϕ_1 and ϕ_2 are displayed. Notice that when the roots are real (in quadrants 1 and 2), the ACF of an $\mathrm{AR}(2)$ model behaves like an $\mathrm{AR}(1)$ model. It is either exponentially decreasing, as in quadrant 1, or alternating, as in quadrant 2. On the other hand, when the roots are complex conjugate pairs, the ACF behaves like a damped sine wave, as in quadrants 3 and 4. In this case, the $\mathrm{AR}(2)$ model displays what is known as pseudoperiodic behavior. □

To summarize, relationships between causality and invertibility of an AR model $\phi(B)Y_t = Z_t$ and an MA model $Y_t = \theta(B)Z_t$ can be represented as follows:

$$Z_t \;\rightarrow\; \boxed{\psi(B) = \phi(B)^{-1}} \;\rightarrow\; Y_t \;; \qquad Y_t \;\rightarrow\; \boxed{\pi(B) = \theta(B)^{-1}} \;\rightarrow\; Z_t.$$

$$\qquad\qquad\quad \underset{\text{Causal}}{} \qquad\qquad\qquad\qquad\qquad \underset{\text{Invertible}}{}$$

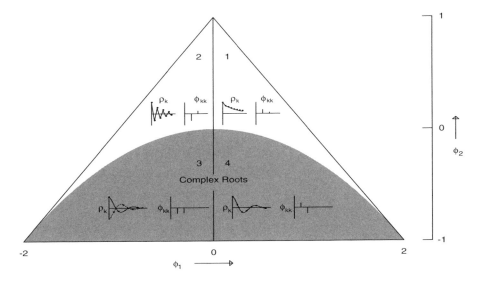

Fig. 3.1   ACF of an AR(2) model.

3.4   ARMA MODELS

Although both the MA and AR models have their own appeal, we may have to use a relatively long AR or long MA model to capture the complex structure of a time series. This may become undesirable since we usually want to fit a parsimonious model; a model with relatively few unknown parameters. To achieve this goal, we can combine the AR and MA parts to form an autoregressive moving average (ARMA) model.

Definition 3.4 $\{Y_t\}$ *is said to be an* **ARMA**(p, q) *process if:*

   (i) $\{Y_t\}$ *is stationary.*

   (ii) For all t, $\phi(B)Y_t = \theta(B)Z_t$, *where* $Z_t \sim \mathrm{WN}(0, \sigma^2)$.

Definition 3.5 $\{Y_t\}$ *is called an* **ARMA**(p, q) **with mean** μ *if* $\{Y_t - \mu\}$ *is an* ARMA(p, q).

   Given the discussions about causality and invertibility, it is prudent to assume that any given ARMA model is causal and invertible. Specifically, let

$$\phi(B)Y_t = \theta(B)Z_t,$$

with

$$\phi(B) = 1 - \phi_1 B - \cdots - \phi_p B^p,$$
$$\theta(B) = 1 - \theta_1 B - \cdots - \theta_q B^q,$$

where $\phi(B)$ and $\theta(B)$ have no common roots, with $\phi(B)$ being causal [i.e., $\phi(B)$ satisfies Theorem 3.2] and $\theta(B)$ being invertible [i.e., $\theta(B)$ satisfies Theorem 3.1]. Under these assumptions, $\{Y_t\}$ is said to be a causal and invertible ARMA(p,q) model. In this case,

$$Y_t = \frac{\theta(B)}{\phi(B)}Z_t = \psi(B)Z_t \;\; ; Z_t \to \boxed{\psi(B) = \frac{\theta(B)}{\phi(B)}} \to Y_t,$$

$$Z_t = \frac{\phi(B)}{\theta(B)}Y_t = \pi(B)Y_t \;\; ; Y_t \to \boxed{\pi(B) = \frac{\phi(B)}{\theta(B)}} \to Z_t.$$

Example 3.4 *Let* $Y_t - \phi Y_{t-1} = Z_t - \theta Z_{t-1}$ *be an* ARMA(1,1) *model with* $\phi = 0.5$ *and* $\theta = 0.3$. *Then*

$$\psi(B) = (1 - 0.3B)\phi^{-1}(B)$$
$$= (1 - 0.3B)(1 + 0.5B + (0.5)^2 B^2 + \cdots)$$
$$= 1 + 0.2B + 0.1B^2 + 0.05B^3 + \cdots$$

Hence, $\psi_i = 0.2 \times (0.5)^{i-1}$, $i = 1, 2, \ldots, \psi_0 = 1$. *Also,* $\pi_i = 0.2 \times (0.3)^{i-1}$, $i = 1, 2, \ldots, \pi_0 = 1$. *Therefore,*

$$\rho(k) = \sum_{i=0}^{\infty} \psi_i \psi_{k+i} \Big/ \sum_{i=0}^{\infty} \psi_i^2$$
$$\sim (0.5)^k. \qquad\qquad \square$$

The usefulness of ARMA models lies in their parsimonious representation. As in the AR and MA cases, properties of ARMA models can usually be characterized by their autocorrelation functions. To this end, a lucid discussion of the various properties of the ACF of simple ARMA models can be found on page 84 of Box, Jenkins, and Reinsel (1994). Further, since the ACF remains unchanged when the process contains a constant mean, adding a constant mean to the expression of an ARMA model would not alter any covariance structure. As a result, discussions of the ACF properties of an ARMA model usually apply to models with zero means.

3.5 ARIMA MODELS

Since we usually process a time series before analyzing it (detrending, for example), it is natural to consider a generalization of ARMA models, the ARIMA model. Let $W_t = (1 - B)^d Y_t$ and suppose that W_t is an ARMA(p, q), $\phi(B)W_t = \theta(B)Z_t$. Then $\phi(B)(1 - B)^d Y_t = \theta(B)Z_t$. The process $\{Y_t\}$ is said to be an ARIMA(p, d, q), autoregressive integrated moving average model.

Usually, d is a small integer (≤ 3). It is prudent to think of differencing as a kind of data transformation.

Example 3.5 *Let $Y_t = \alpha + \beta t + N_t$, so that $(1 - B)Y_t = \beta + Z_t$, where $Z_t = N_t - N_{t-1}$. Thus, $(1 - B)Y_t$ satisfies an MA(1) model, although a non-invertible one. Further, the original process $\{Y_t\}$ is an ARIMA(0,1,1) model, and as such, it is noncausal since it has a unit root.* □

Example 3.6 *Consider an ARIMA(0,1,0), a random walk model:*

$$Y_t = Y_{t-1} + Z_t.$$

If $Y_0 = 0$, then $Y_t = \sum_{i=1}^{t} Z_i$, which implies that var $Y_t = t\sigma^2$. Thus, in addition to being noncausal, this process is also nonstationary, as its variance changes with time. □

As another illustration of ARIMA model, let P_t denote the price of a stock at the end of day t. Define the return on this stock as $r_t = (P_t - P_{t-1})/P_{t-1}$. A simple Taylor's expansion of the log function leads to the following equation:

$$r_t = \frac{P_t - P_{t-1}}{P_{t-1}}$$

$$\cong \log\left(1 + \frac{P_t - P_{t-1}}{P_{t-1}}\right)$$

$$= \log\frac{P_t}{P_{t-1}}$$

$$= \log P_t - \log P_{t-1}.$$

Therefore, if we let $Y_t = \log P_t$, and if we believe that the return on the stock follows a white noise process (i.e., we model $r_t = Z_t$), the derivation above shows that the log of the stock price follows an ARIMA(0,1,0), random walk model. It is because of this that many economists attempt to model the return on an equity (stock, bond, exchange rate etc.) as a random walk model.

In practice, to model possibly nonstationary time series data, we may apply the following steps:

1. Look at the ACF to determine if the data are stationary.

2. If not, process the data, probably by means of differencing.

3. After differencing, fit an ARMA(p, q) model to the differenced data.

Recall that in an ARIMA(p, d, q) model, the process $\{Y_t\}$ satisfies the equation $\phi(B)(1 - B)^d Y_t = \theta(B)Z_t$. It is called *integrated* because of the fact that $\{Y_t\}$ can be recovered by summing (integrating). To see this, consider the following example.

Example 3.7 *Let $\{Y_t\}$ be an* ARIMA(1,1,1) *model that follows*

$$(1 - \phi B)(1 - B)Y_t = Z_t - \theta Z_{t-1}.$$

Then $W_t = (1 - B)Y_t = Y_t - Y_{t-1}$. *Therefore,*

$$\sum_{k=1}^{t} W_k = \sum_{k=1}^{t}(Y_k - Y_{k-1}) = Y_t - Y_0 = Y_t \ \ if \ \ Y_0 = 0.$$

Hence, Y_t is recovered from W_t by summing, hence integrated. The differenced process $\{W_t\}$ satisfies an ARMA(1,1) *model.* □

3.6 SEASONAL ARIMA

Suppose that $\{Y_t\}$ exhibits a seasonal trend, in the sense that $Y_t \sim Y_{t-s} \sim Y_{t-2s} \cdots$. Then Y_t not only depends on $Y_{t-1}, Y_{t-2}, ...,$ but also $Y_{t-s}, Y_{t-2s}, ...$ To model this, consider

$$\phi(B)\Phi_P(B^s)(1 - B)^d(1 - B^s)^D Y_t = \theta(B)\Theta_Q(B^s)Z_t, \qquad (3.10)$$

where

$$\phi(B) = 1 - \phi_1 B - \cdots - \phi_p B^p,$$
$$\theta(B) = 1 - \theta_1 B - \cdots - \theta_q B^q,$$
$$\Phi_p(B^s) = 1 - \Phi_1 B^s - \cdots - \Phi_P B^{s^P},$$
$$\Theta_Q(B^s) = 1 - \Theta_1 B^s - \cdots - \Theta_Q B^{s^Q}.$$

Such $\{Y_t\}$ is usually denoted by SARIMA$(p, d, q) \times (P, D, Q)_s$. Of course, we could expand the right-hand side of (3.10) and express $\{Y_t\}$ in terms of a higher-order ARMA model (see the following example). But we prefer the SARIMA format, due to its natural interpretation.

Example 3.8 *Let us consider the structure of an* SARIMA$(1, 0, 0) \times (0, 1, 1)_{12}$ *time series $\{Y_t\}$, which is expressed as*

$$(1 - \phi B)(1 - B^{12})Y_t = Z_t - \theta Z_{t-12}, (1 - B^{12} - \phi B + \phi B^{13})Y_t = Z_t - \theta Z_{t-12},$$

so that

$$Y_t = Y_{t-12} + \phi\left(Y_{t-1} - Y_{t-13}\right) + Z_t - \theta Z_{t-12}. \qquad (3.11)$$

Notice that Y_t depends on $Y_{t-12}, Y_{t-1}, Y_{t-13}$ as well as Z_{t-12}. If $\{Y_t\}$ repre-sents monthly observations over a number of years, we can tabulate the data *using two-way* ANOVA *as follows*:

	1994	1995	1996
January	Y_1	Y_{13}	Y_{25}
$\vdots$	$\vdots$	$\vdots$	$\vdots$
December	Y_{12}	Y_{24}	Y_{36}

For example, $Y_{26} = f(Y_{25}, Y_{14}, Y_{13}) + \cdots$. *In this case there is an* ARMA *structure for successive months in the same year and an* ARMA *structure for the same month in different years. Note also that according to* (3.11), Y_t *also follows an* ARMA(13,12) *model, with many of the intermediate* AR *and* MA *coefficients being restricted to zeros. Since there is a natural interpretation for an* SARIMA *model, we prefer a* SARIMA *parameterization over a long* ARMA *parameterization whenever a seasonal model is considered.*   □

3.7 EXERCISES

1. Determine which of the following processes are causal and/or invertible:

   (a) $Y_t + 0.2Y_{t-1} - 0.48Y_{t-2} = Z_t$.

   (b) $Y_t + 1.9Y_{t-1} + 0.88Y_{t-2} = Z_t + 0.2Z_{t-1} + 0.7Z_{t-2}$.

   (c) $Y_t + 0.6Y_{t-2} = Z_t + 1.2Z_{t-1}$.

   (d) $Y_t + 1.8Y_{t-1} + 0.81Y_{t-2} = Z_t$.

   (e) $Y_t + 1.6Y_{t-1} = Z_t - 0.4Z_{t-1} + 0.04Z_{t-2}$.

2. Let $\{Y_t : t = 0, \pm 1, \ldots \}$ be the stationary solution of the noncausal AR(1) equation

   $$Y_t = \phi Y_{t-1} + Z_t, \quad |\phi| > 1, \quad \{Z_t\} \sim \text{WN}(0, \sigma^2).$$

   Show that $\{Y_t\}$ also satisfies the causal AR(1) equation

   $$Y_t = \phi^{-1} Y_{t-1} + W_t, \quad \{W_t\} \sim \text{WN}(0, \tilde{\sigma}^2)$$

   for a suitably chosen white noise process $\{W_t\}$. Determine $\tilde{\sigma}^2$.

3. Show that for an MA(2) process with moving average polynomial $\theta(z) = 1 - \theta_1 z - \theta_2 z^2$ to be invertible, the parameters (θ_1, θ_2) must lie in the triangular region determined by the intersection of the three regions

   $$\theta_2 + \theta_1 < 1,$$
   $$\theta_2 - \theta_1 < 1,$$
   $$|\theta_2| < 1.$$

4. Let Y_t be an ARMA(p, q) plus noise time series defined by

$$Y_t = X_t + W_t,$$

where $\{W_t\} \sim \text{WN}(0, \sigma_w^2)$, $\{X_t\}$ is the ARMA(p, q) time series satisfying

$$\phi(B)X_t = \theta(B)Z_t, \quad \{Z_t\} \sim \text{WN}(0, \sigma_z^2),$$

and $E(W_s Z_t) = 0$ for all s and t.

   (a) Show that $\{Y_t\}$ is stationary and find its autocovariance function in terms of σ_w^2 and the ACF of $\{X_t\}$.

   (b) Show that the process $U_t = \phi(B)Y_t$ is r-correlated, where $r = \max(p, q)$, and hence it is an MA(r) process. Conclude that $\{Y_t\}$ is an ARMA(p, r) process.

5. Consider the time series $Y_t = A \sin \omega t + Z_t$, where A is a random variable with mean zero and variance 1, ω is a fixed constant between $(0, \pi)$, and $Z_t \sim \text{WN}(0, \sigma^2)$, which is uncorrelated with the random variable A. Determine if $\{Y_t\}$ is weakly stationary.

6. Suppose that $Y_t = (-1)^t Z$, where Z is a fixed random variable. Give necessary and sufficient condition(s) on Z so that $\{Y_t\}$ will be weakly stationary.

7. Let $Y_t = Z_t - \theta Z_{t-1}$, $Z_t \sim \text{WN}(0, \sigma^2)$.

   (a) Calculate the correlation function $\rho(k)$ of Y_t.

   (b) Suppose that $\rho(1) = 0.4$. What value(s) of θ will give rise to such a value of $\rho(1)$? Which one would you prefer? Give a one-line explanation.

   (c) Instead of an MA(1) model, suppose that Y_t satisfies an infinite MA expression as follows:

$$Y_t = Z_t + C(Z_{t-1} + Z_{t-2} + \cdots), \tag{3.12}$$

   where C is a fixed constant. Show that Y_t is nonstationary.

   (d) If $\{Y_t\}$ in equation (3.12) is differenced (i.e., $X_t = Y_t - Y_{t-1}$), show that X_t is a stationary MA(1) model.

   (e) Find the autocorrelation function of $\{X_t\}$.

8. Consider the time series

$$Y_t = 0.4Y_{t-1} + 0.45Y_{t-2} + Z_t + Z_{t-1} + 0.25Z_{t-2},$$

where $Z_t \sim \text{WN}(0, \sigma^2)$.

(a) Express this equation in terms of the backshift operator B; that is, write it as an equation in B, and determine the order (p, d, q) of this model.

(b) Can you simplify this equation? What is the order after simplification?

(c) Determine if this model is causal and/or invertible.

(d) If the model is causal, find the general form of the coefficients ψ_j's so that $Y_t = \sum_{j=0}^{\infty} \psi_j Z_{t-j}$.

(e) If the model is invertible, find the general form of the coefficients π_j's so that $Z_t = \sum_{j=0}^{\infty} \pi_j Y_{t-j}$.

4

Estimation in the Time Domain

4.1 INTRODUCTION

Let $\{Y_t\}$ be an ARIMA(p, d, q) model that has the form

$$\phi(B)(1 - B)^d(Y_t - \mu) = \theta(B)Z_t, \ Z_t \sim \text{WN}(0, \sigma^2).$$

The unknown parameters in this model are $(\mu, \phi_1, \dots, \phi_p, \theta_1, \dots, \theta_q, \sigma^2)'$ and the unknown orders (p, d, q). We shall discuss the estimation of these parameters from a time-domain perspective. Since the orders of the model, (p, d, q), can be determined (at least roughly) by means of inspecting the sample ACF, let us suppose that the orders (p, d, q) are known for the time being. As in traditional regressions, several statistical procedures are available to estimate these parameters. The first one is the classical method of moments.

4.2 MOMENT ESTIMATORS

The simplest type of estimators are the moment estimates. If $EY_t = \mu$, we simply estimate μ by $\bar{Y} = (1/n) \sum_{t=1}^{n} Y_t$ and proceed to analyze the demeaned series $X_t = Y_t - \bar{Y}$. For the covariance and correlation functions, we may use the same idea to estimate $\gamma(k)$ by

$$C_k = \frac{1}{n} \sum_{t=1}^{n-k} (Y_t - \bar{Y})(Y_{t+k} - \bar{Y}).$$

Similarly, we can estimate $\rho(k)$ by

$$r_k = \frac{C_k}{C_0}.$$

One desirable property of r_k is that it can be shown that when $Y_t \sim \text{WN}(0, 1)$, $r_k \sim \text{AN}(0, 1/n)$. Therefore, a 95% CI of ρ_k is given by $\pm 2/\sqrt{n}$. However, this method becomes unreliable when the lag, k, is big. A rule of thumb is to estimate $\rho(k)$ by r_k for $k < n/3$ or for k no bigger than $O(\log(n))$. Because of the ergodicity assumption, moment estimators are very useful in estimating the mean or the autocovariance structure. Estimations of the specific AR or MA parameters are different, however.

4.3 AUTOREGRESSIVE MODELS

Given the strong resemblance between an $\text{AR}(p)$ model and a regression model, it is not surprising to anticipate that estimation of an $\text{AR}(p)$ model is straightforward. Consider an $\text{AR}(p)$ process

$$Y_t = \phi_1 Y_{t-1} + \cdots + \phi_p Y_{t-p} + Z_t. \tag{4.1}$$

This equation bears a strong resemblance to traditional regression models. Rewriting this equation in the familiar regression expression,

$$Y_t = (\phi_1, \ldots, \phi_p) \begin{pmatrix} Y_{t-1} \\ \vdots \\ Y_{t-p} \end{pmatrix} + Z_t = \boldsymbol{Y}'_{t-1}\boldsymbol{\phi} + Z_t,$$

where $\boldsymbol{\phi} = (\phi_1, \ldots, \phi_p)'$ and $\boldsymbol{Y}_{t-1} = (Y_{t-1}, \ldots, Y_{t-p})'$. The least squares estimate (LSE) of $\boldsymbol{\phi}$ is given by

$$\hat{\boldsymbol{\phi}} = \left(\sum_{t=1}^{n} \boldsymbol{Y}_{t-1}\boldsymbol{Y}'_{t-1} \right)^{-1} \left(\sum_{t=1}^{n} \boldsymbol{Y}_{t-1}Y_t \right).$$

Standard regression analysis can be applied here with slight modifications. Furthermore, if $Z_t \sim \text{N}(0, \sigma^2)$ i.i.d., then $\hat{\boldsymbol{\phi}}$ is also the maximum likelihood estimate (MLE). In the simple case that $p = 1$, $Y_t = \phi Y_{t-1} + Z_t$, we have $\hat{\phi} = \sum_{t=1}^{n} Y_t Y_{t-1} / \sum_{t=1}^{n} Y_{t-1}^2$.

Further, $\hat{Z}_t = Y_t - \boldsymbol{Y}'_{t-1}\hat{\boldsymbol{\phi}}$ is the fitted residual and almost all techniques concerning residual analysis from classical regression can be carried over. Finally, standard asymptotic results such as consistency and asymptotic normality are available.

Theorem 4.1

$$\sqrt{n}\,(\hat{\boldsymbol{\phi}} - \boldsymbol{\phi}) \underset{\mathcal{L}}{\to} \text{N}(0, \sigma^2 \Gamma_p^{-1}),$$

where $\underset{\mathcal{L}}{\rightarrow}$ denotes convergence in distribution of the corresponding random variables as the sample size $n \rightarrow \infty$ and

$$\Gamma_p = E\left(\begin{pmatrix} Y_1 \\ \vdots \\ Y_p \end{pmatrix}(Y_1, \ldots, Y_p)\right) = \begin{pmatrix} \gamma(0) & \gamma(1) & \cdots & \gamma(p-1) \\ * & \gamma(0) & \cdots & \gamma(p-2) \\ \vdots & & & \vdots \\ * & * & \cdots & \gamma(0) \end{pmatrix}.$$

Example 4.1 *For an* AR(1) *model, we have* $\sqrt{n}\,(\hat{\phi} - \phi)\underset{\mathcal{L}}{\rightarrow} N(0, \sigma^2/\gamma(0))$, *with*

$$\gamma(0) = \text{var } Y_t = \phi^2 \text{ var } Y_{t-1} + \sigma^2$$
$$= \phi^2\gamma(0) + \sigma^2.$$

Thus, $\gamma(0) = \sigma^2/(1 - \phi^2)$ [i.e., $\hat{\phi} \sim AN(\phi, (1 - \phi^2)/n)$]. □

From the preceding theorem, usual inference such as constructing approximated confidence intervals or testings for ϕ can be conducted. Alternatively, we can evaluate the Yule–Walker (Y–W) equation via multiplying equation (4.1) by Y_{t-k} and taking expectations,

$$\begin{aligned} \gamma(k) &= \phi_1\gamma(k-1) + \cdots + \phi_p\gamma(k-p), \\ \rho(k) &= \phi_1\rho(k-1) + \cdots + \phi_p\rho(k-p), \quad k = 1, \ldots, p. \end{aligned}$$

In matrix notation, these equations become

$$\begin{pmatrix} \rho(1) \\ \vdots \\ \rho(p) \end{pmatrix} = \begin{pmatrix} \rho(0) & \rho(1) & \cdots & \rho(p-1) \\ \rho(1) & \rho(0) & \cdots & \rho(p-2) \\ \vdots & & & \\ \rho(p-1) & \cdots & & \rho(0) \end{pmatrix}\begin{pmatrix} \phi_1 \\ \vdots \\ \phi_p \end{pmatrix}.$$

Hence, the Yule–Walker estimates are the ϕ such that

$$\hat{\phi} = R^{-1}r = \begin{pmatrix} 1 & r_1 & \cdots & r_{p-1} \\ r_1 & 1 & \cdots & r_{p-2} \\ r_{p-1} & \cdots & \cdots & 1 \end{pmatrix}^{-1}\begin{pmatrix} r_1 \\ \vdots \\ r_p \end{pmatrix}.$$

Again, asymptotic properties of the Y-W estimates can be found. When the sample size n is big and the order p is moderate, computational cost can be enormous for inverting the matrix R. In practice, it would be much more desirable to solve these quantities in real time (i.e., in a recursive on-line manner). The Durbin–Levinson (D–L) algorithm offers such a recursive scheme. We would not pursue the details of this algorithm, but refer the interested reader to the discussion given in Brockwell and Davis (1991). In any case, most of the computer routines, including those of SPLUS programs, use this algorithm to estimate the parameters. Roughly speaking, we can classify the estimation steps as follows:

1. Use the Durbin–Levinson algorithm to evaluate the Yule–Walker estimates.

2. Use the Yule–Walker estimates as initial values to calculate the maximum likelihood estimates (MLE) of the parameters. Details of the MLE are given in Section 4.6.

3. To estimate the standard error in the AR equation, use the estimator

$$\hat{\sigma}^2 = \frac{1}{n} \sum_{t=1}^{n} (Y_t - \hat{Y}_t)^2, \quad \hat{Y}_t = \hat{\phi}_1 Y_{t-1} + \cdots + \hat{\phi}_p Y_{t-p}.$$

4.4 MOVING AVERAGE MODELS

Contrary to the AR model, estimation for an MA model is much more tricky. To illustrate this point, consider the simple MA(1) model $Y_t = Z_t - \theta Z_{t-1}$. Suppose that we intend to use a moment estimator for θ. Then

$$\rho_1 = \frac{-\theta}{1 + \theta^2}$$

and

$$r_1 = \frac{-\hat{\theta}}{1 + \hat{\theta}^2}.$$

Thus,

$$\hat{\theta} = \frac{-1 \pm \sqrt{1 - 4r_1^2}}{2r_1}.$$

This estimator is nonlinear in nature. Such a nonlinearity phenomenon is even more prominent for an MA(q) model. In general, it will be very difficult to express the θ_i's of an MA(q) model as functions of r_i's analytically.

Alternatively, if $| \theta | < 1$, then

$$Z_t = Y_t + \theta Z_{t-1} = Y_t + \theta Y_{t-1} + \theta^2 Y_{t-2} + \cdots .$$

Let $S(\theta) = \sum_{t=1}^{n} Z_t^2$. We can find the θ such that $S(\theta)$ is minimized, where $Z_t = Z_t(\theta)$ implicitly. Note that even in this simple MA(1) case, $S(\theta)$ cannot be minimized analytically. In particular, for given $Y_1, \ldots, Y_n$ and θ, and conditional on $Z_0 = 0$, set

$$\begin{aligned}
Z_1 &= Y_1, \\
Z_2 &= Y_2 + \theta Z_1 = Y_2 + \theta Y_1, \\
&\vdots \\
Z_n &= Y_n + \theta Z_{n-1},
\end{aligned}$$

and compute $S_*(\theta) = \sum_{t=1}^n Z_t^2$ for the given θ, where we use S_* to denote that this quantity is evaluated conditional on the initial value $Z_0 = 0$. In general, we can perform a grid search over $(-1, 1)$ to find the minimum of $S_*(\theta)$ by means of a numerical method, the Gauss–Newton method, say. This is also known as the *conditional least squares* (CLS) *method*. Specifically, consider

$$Z_t(\theta) \cong Z_t(\theta^*) + (\theta - \theta^*)\frac{dZ_t(\theta)}{d\theta}\Big|_{\theta=\theta^*}, \tag{4.2}$$

from an initial point θ^*. Note that this equation is linear in θ, thus $\sum_{t=1}^n Z_t^2(\theta) = S_*(\theta)$ can be minimized analytically to get a new $\theta_{(1)}$. Substitute $\theta_{(1)}$ for θ^* into (4.2) again and iterate this process until it converges. Note that the quantities

$$\frac{dZ_t(\theta^*)}{d\theta} = \theta^*\frac{dZ_{t-1}(\theta^*)}{d\theta} + Z_{t-1}(\theta^*), \ \frac{dZ_0(\theta)}{d\theta} = 0$$

can be evaluated recursively.

For a general MA(q) model, a multivariate Gauss–Newton procedure can be used to minimize $S_*(\boldsymbol{\theta})$ via $Z_t = Y_t + \theta_1 Z_{t-1} + \cdots + \theta_q Z_{t-q}$ such that $Z_0 = Z_{-1} = \cdots = Z_{1-q} = 0$, where $\boldsymbol{\theta} = (\theta_1, \ldots, \theta_q)'$.

4.5   ARMA MODELS

Having seen the intricacies in estimating an MA model, we now discuss the estimation of an ARMA model by means of a simple ARMA(1,1) model.

Example 4.2 *Let* $Y_t - \phi Y_{t-1} = Z_t - \theta Z_{t-1}$. *Conditional on* $Z_0 = 0 = Y_0$, *find* (ϕ, θ) *that minimizes*

$$S_*(\phi, \theta) = \sum_{t=1}^n Z_t^2(\phi, \theta),$$

where

$$Z_t = Y_t - \phi Y_{t-1} + \theta Z_{t-1}. \qquad \qquad \Box$$

For a general ARMA(p, q), we perform a similar procedure to find the estimates by solving a numerical minimization problem. Let $Z_t = Y_t - \phi_1 Y_{t-1} - \cdots - \phi_p Y_{t-p} + \theta_1 Z_{t-1} + \cdots + \theta_q Z_{t-q}$. Compute $Z_t = Z_t(\boldsymbol{\phi}, \boldsymbol{\theta})$, $t = p+1, \ldots, n$ and find the parameters $(\boldsymbol{\phi}, \boldsymbol{\theta})$ that minimize

$$S_*(\boldsymbol{\phi}, \boldsymbol{\theta}) = \sum_{t=p+1}^n Z_t^2(\boldsymbol{\phi}, \boldsymbol{\theta}).$$

For an invertible MA or ARMA model, the initial values of $Y_0 = Y_{-1} = Y_{1-p} = \cdots = Z_0 = \cdots = Z_{1-q} = 0$ have little effect on the final parameter estimates when the sample size is large.

4.6 MAXIMUM LIKELIHOOD ESTIMATES

Previous discussions focussed mainly on least squares procedure, that is, finding an estimate that minimizes some forms of the sum of mean square errors, $\sum Z_t^2(\phi, \theta)$. Another commonly used statistical procedure in this situation is the method of maximum likelihood (ML). In general, the likelihood procedure looks for the parameter value that corresponds most closely with the data observed. To illustrate the likelihood principle, consider the following recapturing example adopted from Feller (1968).

Example 4.3 (Estimation of Population Size) *Suppose that a fund manager wants to launch a financial product in a new region. He wants to know the demand for the product and thus is interested in knowing the population size in this region. The manager decides to conduct a survey and interviews* 1000 *clients at random in the first round. During the interview, each of these clients is issued an identification code. After several months, the manager interviews another round of* 1000 *clients at random, and it is found that* 100 *clients in the second round were interviewed in the first round. What conclusion can be drawn concerning the size of the population in this region? To answer this question, define the following:*

$$n = unknown\ population\ size$$
$$n_1 = number\ of\ people\ interviewed\ in\ the\ first\ round$$
$$r = number\ of\ people\ interviewed\ in\ the\ second\ round$$
$$k = number\ of\ people\ in\ the\ second\ round\ who\ were$$
$$identified\ in\ the\ first\ round$$
$$q_k(n) = probability\ that\ the\ second\ round\ interview\ contains\ exactly$$
$$k\ people\ who\ were\ identified\ in\ the\ first\ round$$

Using the hypergeometric distribution, it is seen that

$$q_k(n) = \frac{\binom{n_1}{k}\binom{n-n_1}{r-k}}{\binom{n}{r}}.$$

In this example, we have $n_1 = r = 1000$ *and* $k = 100$. *If* $n = 1900$, *substituting these numbers into the expression above, we get*

$$q_{100}(n) = \frac{(1000!)^2}{(100!1900!)} = 10^{-430}.$$

With a population size of 1900, *it is virtually impossible that the same* 100 *people interviewed in the first round would be interviewed again during the second round. For any given particular set of* n_1, r, *and* k, *we may want to* *find the value of n that maximizes the probability $q_n(k)$ since for such an n, our observed data would have the greatest probability. This is the key idea of the likelihood principle, and the value $\hat{n}$ is called the* **maximum likelihood estimate***. In other words, for a given set of data, the value $\hat{n}$ is among all possible values of n which is most consistent with the given set of data in terms of maximizing the likelihood. To compute the maximum likelihood, consider the ratio*

$$\frac{q_k(n)}{q_k(n-1)} = \frac{(n-n_1)(n-r)}{(n-n_1-r+k)n}.$$

It can easily be seen that

$$\frac{q_k(n)}{q_k(n-1)} \begin{cases} > 1, & \text{if } nk < n_1 r, \\ < 1, & \text{if } nk > n_1 r. \end{cases}$$

This means that when n increases, the sequence $q_k(n)$ first increases and then decreases. It reaches its maximum when n is the largest integer before the number $n_1 r/k$. Thus, $\hat{n} \sim n_1 r/k$. In this example, the maximum likelihood estimate of the population size $\hat{n} = 10,000$.   □

Recall that if $X_1, \ldots, X_n$ are i.i.d. random variables with a common probability density function $f(x, \theta)$, the likelihood equation is defined as

$$L(\boldsymbol{x}, \theta) = \prod_{i=1}^{n} f(x_i, \theta). \tag{4.3}$$

The maximum likelihood estimate $\hat{\theta}$, of θ is obtained by finding the value of θ that maximizes (4.3). In other words, we would like to find the value of the unknown parameter that maximizes the likelihood (probability) that it happens for a given set of observations $x_1, \ldots, x_n$. Notice that this is a very general procedure and the i.i.d. assumption can be relaxed. For a time series problem, this idea can still be applied and we shall illustrate this idea through an example.

Consider an AR(1) model $Y_t = \phi Y_{t-1} + Z_t$, $Z_t \sim N(0, \sigma^2)$, i.i.d. The joint probability density function (pdf) of $(Z_2, \ldots, Z_n)$ is given by

$$f(Z_2, \ldots, Z_n) = \left(\frac{1}{2\pi\sigma^2}\right)^{(n-1)/2} \exp\left(\frac{-1}{2\sigma^2} \sum_{t=2}^{n} Z_t^2\right),$$

where $Y_2 = \phi Y_1 + Z_2, \ldots, Y_n = \phi Y_{n-1} + Z_n$. By means of the transformation of variables method and the fact that the determinant of the Jacobian of this transformation is 1, the joint pdf of $Y_2, \ldots, Y_n$ conditional on Y_1 is given by

$$f(Y_2, \ldots, Y_n \mid Y_1) = \left(\frac{1}{2\pi\sigma^2}\right)^{(n-1)/2} \exp\left[\frac{-1}{2\sigma^2} \sum_{t=2}^{n} (Y_t - \phi Y_{t-1})^2\right].$$

Recall that $Y_1 \sim N\left(0, \sigma^2/(1 - \phi^2)\right)$ if $Z_0, Z_{-1}, Z_{-2}, \ldots$ are i.i.d. $N(0, \sigma^2)$. Since

$$f(Y_1) = \left(\frac{1}{2\pi\sigma^2}\right)^{1/2} \sqrt{1 - \phi^2} \, \exp\left[\frac{-1}{2\sigma^2}(1 - \phi^2)Y_1^2\right],$$

the likelihood function is given by

$$
\begin{aligned}
L(\phi) &= f(Y_1, \ldots, Y_n) \\
&= f(Y_2, \ldots, Y_n \mid Y_1)f(Y_1) \\
&= \left(\frac{1}{2\pi\sigma^2}\right)^{n/2} \sqrt{1 - \phi^2} \exp\left\{\frac{-1}{2\sigma^2}\left[\sum_{t=2}^{n}(Y_t - \phi Y_{t-1})^2 + (1 - \phi^2)Y_1^2\right]\right\} \\
&= \left(\frac{1}{2\pi\sigma^2}\right)^{n/2} \sqrt{1 - \phi^2} \exp\left[\frac{-1}{2\sigma^2}S(\phi)\right],
\end{aligned}
$$

where $S(\phi) = \sum_{t=1}^{n}(Y_t - \phi Y_{t-1})^2 + (1 - \phi)^2 Y_1^2$. Therefore, the log-likelihood function becomes

$$
\begin{aligned}
\lambda(\phi, \sigma^2) &= \log L(\phi) \\
&= \frac{n}{2}\log 2\pi - \frac{n}{2}\log \sigma^2 + \frac{1}{2}\log(1 - \phi^2) - \frac{1}{2\sigma^2}S(\phi). \quad (4.4)
\end{aligned}
$$

For a given ϕ, the log-likelihood function λ can be maximized with respect to σ^2 by setting $\partial\lambda/\partial\sigma^2 = 0$. Solving this equation leads to $\hat{\sigma}^2 = S(\hat{\phi})/n$. Further, since

$$S(\phi) = S_*(\phi) + (1 - \phi^2)Y_1^2,$$

where $S_*(\phi) = \sum_{t=2}^{n}(Y_t - \phi Y_{t-1})^2$, for moderate to large n, the second term in this expression is negligible with respect to $S_*(\phi)$ and $S(\phi) \cong S_*(\phi)$. Hence for a large sample size n, the value of ϕ that minimizes $S(\phi)$ and $S_*(\phi)$ are similar, and minimizing the unconditional sum of squares $S(\phi)$ over ϕ is tantamount to minimizing the conditional sum of squares $S_*(\phi)$.

Similarly, when maximizing the full log-likelihood λ, the dominating factor is $S(\phi)$ for large n since $\frac{1}{2}\log(1 - \phi^2)$ does not involve n, except in the case when the minimizer occurs at ϕ near 1. This is another reason why precautions need to be taken when dealing with a nonstationary or nearly nonstationary AR(1) model (i.e., when $\phi \cong 1$).

As a compromise between CLS and MLE, we find the estimator such that $S(\phi)$ is minimized. This is known as the *unconditional LSE*. Starting with the most precise method and continuing in decreasing order of precision, we can summarize the various methods of estimating an AR(1) model as follows:

1. *Exact likelihood method.* Find ϕ such that $\lambda(\phi)$ is maximized. This is usually nonlinear and requires numerical routines.

2. *Unconditional least squares.* Find ϕ such that $S(\phi)$ is minimized. Again, nonlinearity dictates the use of numerical routines.

3. *Conditional least squares.* Find ϕ such that $S_*(\phi)$ is minimized. This is the simplest case since $\hat{\phi}$ can be solved analytically.

Notice that all three procedures are asymptotically equivalent. In general, for an ARMA(p,q) model, we can use any one of these methods in a similar manner. To find the unconditional least squares, we need to find ϕ such that $S(\phi)$ is minimized. As in the case of CLS, the explicit form for such an estimator is usually complicated for a general ARMA model.

Finally, similar to Theorem 4.1, there is also a central limit theorem for the MLE $\hat{\phi}$ of an ARMA(p,q) model. We refer readers to Brockwell and Davis (1991) for a general statement about this result.

4.7 PARTIAL ACF

Recall that we can identify the order of an MA model by inspecting its ACF. We now introduce a similar device to identify the order of an AR model. For any given $k > 1$, let a collection of a stationary time series, $\{Y_{t-k}, Y_{t-k+1}, \ldots, Y_{t-1}, Y_t\}$ be given. Consider predicting Y_t linearly based on $\{Y_{t-k+1}, \ldots, Y_{t-1}\}$. Denote this predictor by $\hat{Y}_t$. In a mathematical context, this means projecting Y_t linearly onto the space spanned by the random variables $\{Y_{t-k+1}, \ldots, Y_{t-1}\}$; see, for example, Brockwell and Davis (1991) for further discussions using the Hilbert space formualtion. In other words, $\hat{Y}_t = P_{\overline{sp}\{Y_{t-k+1}, \ldots, Y_{t-1}\}} Y_t$. Then

$$\hat{Y}_t = \beta_1 Y_{t-1} + \cdots + \beta_{k-1} Y_{t-k+1}$$

and

$$\hat{Z}_t = Y_t - \beta_1 Y_{t-1} - \cdots - \beta_{k-1} Y_{t-k+1}.$$

Allowing time to travel backward, consider "predicting" Y_{t-k} linearly based on $\{Y_{t-k+1}, \ldots, Y_{t-1}\}$; we have

$$\hat{Y}_{t-k} = \beta_1 Y_{t-k+1} + \cdots + \beta_{k-1} Y_{t-1}$$

and

$$\hat{Z}_{t-k} = Y_{t-k} - \beta_1 Y_{t-k+1} - \cdots - \beta_{k-1} Y_{t-1}.$$

The reason that the coefficients of these two predictions end up being identical is a consequence of the stationarity of the time series. Since these coefficients can be determined through the projection equation, which in turn depends on the covariance structure of the time series, it can be shown that under the stationarity assumption, the coefficients of these two predictions are exactly the same.

Now, consider the correlation coefficient between these two sets of residuals, $\hat{Z}_t$ and $\hat{Z}_{t-k}$. This is defined as the partial autocorrelation coefficient (PACF) of order k. It captures the relationship between Y_t and Y_{t-k} that is not explained by the predictors $Y_{t-k+1}, \ldots, Y_{t-1}$ (i.e., after regressing Y_t and Y_{t-k} on the intermediate observations $Y_{t-k+1}, \ldots, Y_{t-1}$). The PACF is usually denoted by either $\alpha(k)$ or ϕ_{kk}. Note that $\phi_{kk} = \text{corr}\,(\hat{Z}_{t-k}, \hat{Z}_t)$ with $\phi_{11} = \rho(1)$. Formally, we define

Definition 4.1 *The* **PACF** *of a stationary time series is defined as*

$$\phi_{11} = \rho(1),$$

$$\phi_{kk} = \text{corr}\,(Y_{k+1} - P_{\bar{sp}\{Y_2,\cdots,Y_k\}}Y_{k+1}, Y_1 - P_{\bar{sp}\{Y_2,\cdots,Y_k\}}Y_1), \quad k \geq 2,$$

where $P_{\bar{sp}\{Y_2,\ldots,Y_k\}}Y$ denotes the projection of the random variable Y onto the closed linear subspace spanned by the random variables $\{Y_2, \ldots, Y_k\}$.

Example 4.4 *Consider the simple case where we only have $\{Y_1, Y_2, Y_3\}$. Then*

$$\hat{Y}_3 = \rho_1 Y_2 \quad \text{and} \quad \hat{Y}_1 = \rho_1 Y_2.$$

Therefore,

$$\phi_{22} = \text{corr}\,(Y_3 - \rho_1 Y_2, Y_1 - \rho_1 Y_2)$$
$$= \frac{\text{cov}\,(Y_3 - \rho_1 Y_2, Y_1 - \rho_1 Y_2)}{[\text{var}\,(Y_3 - \rho_1 Y_2)\,\text{var}\,(Y_1 - \rho_1 Y_2)]^{1/2}}. \tag{4.5}$$

Now,

$$\text{cov}\,(Y_3 - \rho_1 Y_2, Y_1 - \rho_1 Y_2) = \gamma_2 - \rho_1 \gamma_1 - \rho_1 \gamma_1 + \rho_1^2 \gamma_0$$
$$= \gamma_0(\rho_2 - 2\rho_1^2 + \rho_1^2)$$
$$= \gamma_0(\rho_2 - \rho_1^2)$$

and

$$\text{var}\,(Y_3 - \rho_1 Y_2) = \text{var}\,Y_3 - 2\rho_1\,\text{cov}\,(Y_3, Y_2) + \rho_1^2\,\text{var}\,Y_2$$
$$= \gamma_0(1 - 2\rho_1 \gamma_1 + \rho_1^2)$$
$$= \gamma_0(1 - \rho_1^2).$$

Substituting these back to (4.5), we obtain

$$\phi_{22} = \frac{\rho_2 - \rho_1^2}{1 - \rho_1^2}.$$

As a consequence, for an $\text{AR}(1)$ model, $\rho_k = \phi^k, \phi_{22} = 0$. □

In general, for an $\text{AR}(p)$ model, it can easily be seen that for $n > p$, $\hat{Y}_n = \phi_1 Y_{n-1} + \cdots + \phi_p Y_{n-p}$. On the other hand, $\hat{Y}_1 = h(Y_2, \ldots, Y_{n-1})$ for some function h. Thus, $\text{cov}\,(\hat{Z}_n, \hat{Z}_1) = \text{cov}\,(Y_n - \hat{Y}_n, Y_1 - \hat{Y}_1) = \text{cov}\,(Z_n, Y_1 -$

$h(Y_2, \dots, Y_{n-1})$). Since $\{Y_t\}$ is an AR(p) process, the random variable $Y_1 - h(Y_2, \dots, Y_{n-1})$ can be written as $g(Z_1, Z_2, \dots, Z_{n-1})$ for some function g. As Z_n is uncorrelated with $Z_1, \dots, Z_{n-1}$, it is uncorrelated with $g(Z_1, \dots, Z_n)$, and consequently,

$$
\begin{aligned}
\mathrm{cov}\,(\hat{Z}_n, \hat{Z}_1) &= \mathrm{cov}\,(Z_n, Y_1 - h(Y_2, \dots, Y_{n-1})) \\
&= \mathrm{cov}\,(Z_n, g(Z_1, \dots, Z_n)) \\
&= 0.
\end{aligned}
$$

Hence, we have the following result.

Theorem 4.2 *For an* AR(p), *$\phi_{kk} = 0$ for $k > p$.*

In practice, we replace the ρ's by r's to calculate the sample PACF $\hat{\phi}_{kk}$. Equivalently, it can be shown that from the Yule–Walker equations, ϕ_{kj}, $j = 1, \dots, k$; $k = 1, 2, \dots$ can be obtained by solving the equations

$$
\begin{pmatrix}
\rho(0) & \rho(1) & \cdots & \rho(k-1) \\
\rho(1) & \rho(0) & & \rho(k-2) \\
& & \cdots & \\
& & & \rho(0)
\end{pmatrix}
\begin{pmatrix}
\phi_{k1} \\
\vdots \\
\phi_{kk}
\end{pmatrix}
=
\begin{pmatrix}
\rho(1) \\
\vdots \\
\rho(k)
\end{pmatrix}.
$$

Therefore, the PACF can be obtained by solving for the last coefficient in the Yule–Walker equation. In particular, the sample PACF is obtained by replacing the $\rho(k)$'s by the r_k's in the Yule–Walker equation. The proof of this fact can be found in Brockwell and Davis (1991). Similar to the situation of the sample ACF r_k, there is also a central limit theorem for the PACF $\hat{\phi}_{kk}$.

Theorem 4.3 *For an* AR(p) *model,* $\sqrt{n}\,\hat{\phi}_{kk} \sim \mathrm{AN}(0, 1)$ *for $k > p$.*

As a result, we can use the sample PACF, $\hat{\phi}_{kk}$, to identify the order of an AR model in the same way that we use the sample ACF, r_k, to identify the order of an MA model. It should be noted, however, that there is no clear pattern for the sample PACF of an MA model.

4.8 ORDER SELECTIONS*

Although we can use the ACF and the PACF to determine the tentative orders p and q, it seems more desirable to have a systematic order selection criterion for a general ARMA model. There are two commonly used methods, the FPE (final prediction error) and the AIC (Akaike's information criterion). Both methods are best illustrated through an AR model.

---

*Throughout this book, an asterisk indicates a technical section that may be browsed casually without interrupting the flow of ideas.

To consider the FPE, suppose that we are given a realization $X = (X_1, \ldots, X_n)$ of an AR(p) model ($p < n$). Let $Y = (Y_1, \ldots, Y_n)$ be an independent realization of the same process. Let $\hat{\phi}_1, \ldots, \hat{\phi}_p$ be the MLE based on X. Consider the one-step-ahead predictor of Y_{n+1} using these $\hat{\phi}$'s, that is,

$$\hat{Y}_{n+1} = \hat{\phi}_1 Y_n + \cdots + \hat{\phi}_p Y_{n-p}.$$

Then the MSE becomes

$$
\begin{aligned}
\text{MSE} &= E(Y_{n+1} - \hat{Y}_{n+1})^2 \\
&= E(Y_{n+1} - \hat{\phi}_1 Y_n - \cdots - \hat{\phi}_p Y_{n-p})^2 \\
&= E(Y_{n+1} - \phi_1 Y_n - \cdots - \phi_p Y_{n-p} \\
&\quad - (\hat{\phi}_1 - \phi_1)Y_n - \cdots - (\hat{\phi}_p - \phi_p)Y_{n-p})^2 \\
&= \sigma^2 + E\left((\hat{\phi} - \phi)'(Y_{n+1-i}Y_{n+1-j})^p_{i,j=1}(\hat{\phi} - \phi) \right),
\end{aligned}
\tag{4.6}
$$

where $\hat{\phi} = (\hat{\phi}_1, \ldots, \hat{\phi}_p)'$ and $\phi = (\phi_1, \ldots, \phi_p)'$.

Because X and Y are independent, the second term in equation (4.6) can be written as the product of the expected values, so that

$$\text{MSE} = \sigma^2 + E((\hat{\phi} - \phi)'\Gamma_p(\hat{\phi} - \phi)),$$

with $\Gamma_p = E(Y_i Y_j)^p_{i,j=1}$. Recall that $n^{1/2}(\hat{\phi} - \phi) \sim \text{AN}(0, \sigma^2 \Gamma_p^{-1})$. Using this fact, we have

$$\frac{\sqrt{n}}{\sigma}(\hat{\phi} - \phi)'\Gamma_p^{1/2}\Gamma_p^{1/2}(\hat{\phi} - \phi)\frac{\sqrt{n}}{\sigma} \sim \chi_p^2.$$

Hence,

$$\text{MSE} \cong \sigma^2 + \frac{\sigma^2 p}{n} = \sigma^2 \left(1 + \frac{p}{n}\right).$$

Further, if $\hat{\sigma}^2$ is the MLE for σ^2, then $n\hat{\sigma}^2/\sigma^2 \sim \chi_{n-p}^2$. Replacing σ^2 by $n\hat{\sigma}^2/(n-p)$ in the expression above, we have

$$\text{FPE} = \hat{\sigma}^2 \left(\frac{n+p}{n-p}\right).
\tag{4.7}$$

Therefore, we would like to find a p such that the FPE is minimized. Note that the right-hand side of (4.7) consists of two quantities: $\hat{\sigma}^2$ and $(n+p)/(n-p)$. We can interpret this equation as follows: When p increases, $\hat{\sigma}^2$ decreases but the quantity $(n+p)/(n-p)$ increases. Thus, minimizing the FPE criterion aims at striking a compromise between these two quantities, and the term $(n+p)/(n-p)$ plays the role of a penalty term when the model complexity p is increased.

A second commonly used criterion is Akaike's information criterion (AIC), based on the Kullback–Leibler information index. Roughly speaking, this index quantifies a metric between two competing models. The key idea is to define an index and to select a model that minimizes this index.

Given an ARMA(p, q) with realization $\{X_1, \dots, X_n\}$, let $\hat{\boldsymbol{\beta}} = (\hat{\boldsymbol{\phi}}, \hat{\boldsymbol{\theta}})$ and $\hat{\sigma}^2$ be MLEs based on $\{X_1, \dots, X_n\}$. Further, assume the X's to be normally distributed and let $\{Y_1, \dots, Y_n\}$ be an independent realization of the same process. Consider the likelihood (log-likelihood) function

$$L(\boldsymbol{\beta}, \sigma^2) = \left(\frac{1}{2\pi\sigma^2}\right)^{n/2} \exp\left[-\frac{1}{2\sigma^2}\sum_{t=1}^{n}(X_t - \phi_1 X_{t-1} - \cdots - \theta_q Z_{t-q})^2\right],$$

$$-2 \log L(\boldsymbol{\beta}, \sigma^2) = n\log 2\pi + n\log \sigma^2 + S_x(\boldsymbol{\beta})/\sigma^2,$$

$$-2 \log L_x(\hat{\boldsymbol{\beta}}, \hat{\sigma}^2) = n\log 2\pi + n\log \hat{\sigma}^2 + S_x(\hat{\boldsymbol{\beta}})/\hat{\sigma}^2$$

$$= n\log 2\pi + n\log \hat{\sigma}^2 + n.$$

Hence,

$$-2\log L_y(\hat{\boldsymbol{\beta}}, \hat{\sigma}^2) = n\log 2\pi + n\log \hat{\sigma}^2 + S_y(\hat{\boldsymbol{\beta}})/\hat{\sigma}^2$$

$$= -2\log L_x(\hat{\boldsymbol{\beta}}, \hat{\sigma}^2) + S_y(\hat{\boldsymbol{\beta}})/\hat{\sigma}^2 - n.$$

Taking expectations yields

$$E(-2\log L_y(\hat{\boldsymbol{\beta}}, \hat{\sigma}^2)) = E(-2\log L_x(\hat{\boldsymbol{\beta}}, \hat{\sigma}^2)) + E(S_y(\hat{\boldsymbol{\beta}})/\hat{\sigma}^2) - n$$

$$= E(\text{Kullback–Leibler index}).$$

To evaluate this expectation, we make use of the following facts. Their proofs can be found in Brockwell and Davis (1991).

1.

$$S_y(\hat{\boldsymbol{\beta}}) \cong S_y(\boldsymbol{\beta}) + (\hat{\boldsymbol{\beta}} - \boldsymbol{\beta})\frac{\partial S_y(\boldsymbol{\beta})}{\partial \boldsymbol{\beta}} + \frac{1}{2}(\hat{\boldsymbol{\beta}} - \boldsymbol{\beta})'\frac{\partial^2 S_y(\boldsymbol{\beta})}{\partial \beta_i\, \partial \beta_j}(\hat{\boldsymbol{\beta}} - \boldsymbol{\beta}).$$

Therefore,

$$E(S_y(\hat{\boldsymbol{\beta}})) \cong ES_y(\boldsymbol{\beta}) + \sigma^2 E((\hat{\boldsymbol{\beta}} - \boldsymbol{\beta})'V^{-1}(\hat{\boldsymbol{\beta}} - \boldsymbol{\beta}))$$

as

$$\hat{\boldsymbol{\beta}} - \boldsymbol{\beta} \sim \text{AN}(0, n^{-1}V) \quad \text{and} \quad \frac{1}{2}\frac{\partial^2 S_y}{\partial \beta_i\, \partial \beta_j} \xrightarrow{P} \sigma^2 V^{-1}.$$

Since

$$E\left(\frac{S_y(\boldsymbol{\beta})}{n}\right) = \sigma^2, \quad \text{we have} \quad \frac{S_y(\boldsymbol{\beta})}{n} \to \sigma^2.$$

Further, since $\hat{\boldsymbol{\beta}} - \boldsymbol{\beta}$ has covariance matrix V, $E((\hat{\boldsymbol{\beta}} - \boldsymbol{\beta})'V^{-1}(\hat{\boldsymbol{\beta}} - \boldsymbol{\beta})) = p + q$. Combining all these facts, we have

$$E(S_y(\hat{\boldsymbol{\beta}})) \cong \sigma^2(n + p + q).$$

2. Since $n \, \hat{\sigma}^2 = S_x(\hat{\boldsymbol{\beta}}) \sim \sigma^2 \chi^2{}_{n-p-q}$, which is independent of $\hat{\boldsymbol{\beta}}$,

$$E\left(\frac{S_y(\hat{\boldsymbol{\beta}})}{\hat{\sigma}^2}\right) \cong \frac{E(S_y(\hat{\boldsymbol{\beta}}))}{E(\hat{\sigma}^2)} \quad \text{(because } X \text{ is independent of } Y)$$

$$= \frac{\sigma^2(n+p+q)}{\sigma^2(n-p-q)/n},$$

$$E\left(\frac{S_y(\hat{\boldsymbol{\beta}})}{\hat{\sigma}^2}\right) - n \cong n[(n+p+q)/(n-p-q) - 1]$$

$$= 2n(p+q)/(n-p-q).$$

3. Therefore, the K-L index can be approximated by

$$-2\log L_x(\hat{\boldsymbol{\beta}}, \hat{\sigma}^2) + \frac{2(p+q)n}{n-p-q} = \text{AICC}$$

[i.e., $E(\text{AICC}(\boldsymbol{\beta})) \cong E(\text{K-L index})$].

Formally, we have the following definition for Akaike's information criterion corrected (AICC) and Akaike's information criterion (AIC).

Definition 4.2

$$\text{AICC}(\boldsymbol{\beta}) = -2\log L_x\left(\boldsymbol{\beta}, \frac{S_x(\boldsymbol{\beta})}{n}\right) + \frac{2(p+q+1)n}{n-p-q-2}, \tag{4.8}$$

$$\text{AIC}(\boldsymbol{\beta}) = -2\log L_x\left(\boldsymbol{\beta}, \frac{S_x(\boldsymbol{\beta})}{n}\right) + 2(p+q+1). \tag{4.9}$$

To use these criteria, fit models with orders p and q such that either AICC or AIC is minimized. Since we never know the true order for sure, in using AIC we should not choose a model based on the minimal value of AIC solely. Within a class of competing models, a minimum value of $p \pm C$ may still be legitimate when other factors, such as parsimony and whiteness of residuals, are taken into consideration.

Another commonly used order selection criterion is the Bayesian information criterion (BIC). It attempts to correct the overfitting nature of the AIC. For a given ARMA(p, q) series $\{X_t\}$, it is defined as follows.

Definition 4.3

$$\text{BIC}(\boldsymbol{\beta}) = (n-p-q)\log[n\hat{\sigma}^2/(n-p-q)] + n(1+\log\sqrt{2\pi})$$

$$+ (p+q)\log\left[\left(\sum_{i=1}^{n} X_i^2 - n\hat{\sigma}^2\right)\Big/(p+q)\right], \tag{4.10}$$

where $\hat{\sigma}^2$ is the maximum likelihood estimate of the white noise sequence.

4.9   RESIDUAL ANALYSIS

After a model has been fitted, the next step to check the model is to perform a residual analysis. Specifically, let the residuals be denoted by $\hat{Z}_t = Y_t - \hat{Y}_t$. Then perform the following steps:

1. Make a time series plot of $\hat{Z}_t$.

2. Plot the ACF of $\hat{Z}_t$.

3. Under the null hypothesis that $Z_t \sim \mathrm{WN}(0,1)$, it can be shown that the r_j (of Z_t) $\sim \mathrm{N}(0, 1/n)$. As a rule of thumb, one can use $\pm 2/\sqrt{n}$ to determine if the observed r_j is significantly different from zero. But remember that only short lags of r_j are meaningful.

4. Instead of looking at the r_j's individually, consider the sum

$$Q = n(n+2) \sum_{j=1}^{h} \hat{r}_Z^2(j)/(n-j), \qquad (4.11)$$

where $\hat{r}_Z(j)$ is the sample correlation of $\hat{Z}_t$ at lag j. This is known as *Portmanteau statistics*. It pools together the information of r_j's for several lags. To use Portmanteau statistics, we need the following result regarding its asymptotic distribution when the Z_t are white noise.

Theorem 4.4 *Let $\{Z_t\} \sim \mathrm{WN}(0,1)$ and Q be the Portmanteau statistic defined in (4.11). Then $Q \rightarrow \chi^2_{(n-p-q)}$ as $n \rightarrow \infty$.*

(i) In practice, h is chosen between 15 and 30.

(ii) This procedure needs large n, say $n \geq 100$.

(iii) The power of this statistic can be poor.

4.10   MODEL BUILDING

Model building can be classified into three stages:

1. Model specification (choosing ARIMA)

2. Model identification (estimation)

3. Model checking (diagnostic)

Corresponding to each of these stages, we have the following procedures.

1. ARIMA, trend + seasonality, and so on.

2. LSE, CLSE, MLE, and so on.

3. Residual analysis, Portmanteau statistics, and so on.

After step 3, remodel the data and return to step 1. This recursion may have to be repeated several times successively before a satisfactory model is attained.

4.11   EXERCISES

1. Consider the conditional least squares procedure for an MA(1) model with $Z_0 = 0$,

$$Y_t = Z_t - \theta Z_{t-1}. \tag{4.12}$$

   (a) For a given initial value θ_0, express the conditional sum of squares $S_*(\theta) = \sum Z_t^2(\theta)$ in terms of Y.

   (b) Perform a Gauss–Newton search to find the value of θ that minimizes $S_*(\theta)$. Using equation (4.2), demonstrate how you would proceed to solve for θ_1 analytically in the first iteration. To solve this, you may assume that the values of $dZ_t(\theta)/d\theta$ are known for the time being.

   (c) Using the defining equation (4.12) of the MA(1) model, show how to compute the derivatives of $dZ_t(\theta)/d\theta$ recursively.

   (d) Once you obtained θ_1, you can repeat the steps above to get a new θ_2 in the same manner and repeat the procedure until it converges. This is the spirit of the Gauss–Newton search.

2. Consider an AR(1) model

$$Y_t = \phi Y_{t-1} + Z_t, \quad Z_t \sim N(0, \sigma^2), \text{i.i.d.}$$

   (a) Verify that the log-likelihood function $\lambda(\phi, \sigma^2)$ is given by equation (4.4).

   (b) Derive that the maximum likelihood estimates of ϕ and σ^2 are given by

$$\hat{\sigma}^2 = S(\hat{\phi})/n,$$

   where

$$S(\hat{\phi}) = \sum_{t=2}^{n}(Y_t - \hat{\phi}Y_{t-1})^2 + (1 - \hat{\phi}^2)Y_1^2,$$

   and $\hat{\phi}$ is the value of ϕ that minimizes the function

$$l(\phi) = -\log(n^{-1}S(\phi)) - n^{-1}\log(1 - \phi^2).$$

3. For an AR(1) process the sample autocorrelation $\hat{\rho}(1)$ is $\text{AN}(\phi, (1 - \phi^2)/n)$. Show that $\sqrt{n} \, (\hat{\rho}(1) - \phi)/(1 - \rho^2(1))^{1/2}$ is $\text{AN}(0, 1)$. If a sample of size 100 from an AR(1) process gives $\hat{\rho}(1) = 0.638$, construct a 95% confidence interval for ϕ. Are the data consistent with the hypothesis that $\phi = 0.7$?

4. Consider the AR(2) process $\{Y_t\}$ satisfying

$$Y_t - \phi Y_{t-1} - \phi^2 Y_{t-2} = Z_t,$$

   where $\{Z_t\} \sim \text{WN}(0, \sigma^2)$.

   (a) For what value of ϕ is this a causal process?

   (b) The following sample moments were computed after observing $Y_1, \dots, Y_{200}$:

   $$\hat{\gamma}(0) = 6.06, \ \hat{\rho}(1) = 0.687, \ \hat{\rho}(2) = 0.610.$$

   Find estimates of ϕ and σ^2 by solving the Yule–Walker equations. If you find more than one solution, choose the one that is causal.

5. Let $\{Y_t\}$ be an ARMA(1,1) time series satisfying

$$Y_t = \phi Y_{t-1} + Z_t - \theta Z_{t-1},$$

   where $\{Z_t\} \sim \text{WN}(0, \sigma^2)$.

   (a) Solve for $\gamma(0)$ and $\gamma(1)$ in terms of ϕ and θ.

   (b) It is given that the MLE

   $$\begin{pmatrix} \hat{\phi} \\ \hat{\theta} \end{pmatrix} \ \text{of} \ \begin{pmatrix} \phi \\ \theta \end{pmatrix}$$

   satisfies the following theorem:

   $$\begin{pmatrix} \hat{\phi} \\ \hat{\theta} \end{pmatrix} \sim \text{AN} \left(\begin{pmatrix} \phi \\ \theta \end{pmatrix}, \frac{\sigma^2}{n} \Gamma^{-1} \right),$$

   where

   $$\Gamma^{-1} = \frac{1 - \phi\theta}{(\phi - \theta)^2} \begin{pmatrix} (1 - \phi^2)(1 - \phi\theta) & -(1 - \theta^2)(1 - \phi^2) \\ -(1 - \theta^2)(1 - \phi^2) & (1 - \theta^2)(1 - \phi\theta) \end{pmatrix}.$$

   Instead of an ARMA(1,1) model, assume that the data are actually generated from an AR(1) model $Y_t = \phi Y_{t-1} + Z_t$. It is known in this case that the MLE $\hat{\phi} \sim \text{AN}(\phi, (\sigma^2/n)(1 - \phi^2))$. Explain what happens to the standard error of the MLE $\hat{\phi}$ if we overfit the AR(1) model by the ARMA(1,1) model.

6. Consider an MA(1) model satistfying

$$Y_t = Z_t - \theta Z_{t-1},$$

where $\{Z_t\} \sim \mathrm{WN}(0, 1)$. It is given that the ACF of $\{Y_t\}$ is given by

$$\gamma(0) = (1 + \theta^2), \ \gamma(1) = -\theta, \ \gamma(k) = 0 \quad \text{for } |k| > 1.$$

Suppose that we have three data points Y_1, Y_2, and Y_3.

(a) Find the coefficient b so that the estimator $\hat{Y}_3 = bY_2$ of Y_3 minimizes

$$E(Y_3 - \hat{Y}_3)^2 = E(Y_3 - bY_2)^2.$$

(b) Similarly, with the value of b given in part (a), let $\hat{Y}_1 = bY_2$ be an estimator of Y_1. Evaluate

$$\mathrm{cov}(Y_3 - \hat{Y}_3, Y_1 - \hat{Y}_1).$$

(c) Hence, deduce that the partial ACF at lag 2 for $\{Y_t\}$ is given by

$$\phi_{22} = -\frac{\theta^2}{1 + \theta^2 + \theta^4}.$$

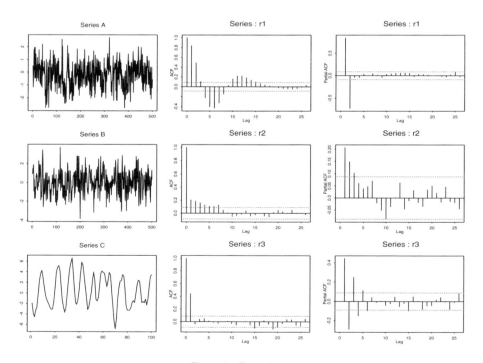

Fig. 4.1 Exercise 7.

(d) Show that for the MA(1) model $\{Y_t\}$,

$$\phi_{kk} = -\frac{\theta^k(1-\theta^2)}{1-\theta^{2(k+1)}} \quad \text{for } k \geq 2.$$

7. Let $\{Z_t\}$ be i.i.d. N(0,1) random variables. We generate three time series according to models (a), (b), and (c) as follows:

(a) $Y_t = 0.85Y_{t-1} + Z_t - 0.7Z_{t-1}$.

(b) $Y_t = Z_t + 0.7Z_{t-1}$.

(c) $Y_t = 1.5Y_{t-1} - 0.75Y_{t-2} + Z_t$.

In Figure 4.1, the first plot in the first column is the time series plot of the series (a), the second plot in the first column is the time series plot of the series (b), and the third plot in the first column is the time series plot of the series (c). The second and third columns of Figure 4.1 are the ACF and PACF of either series (a), (b), or (c). Note that within the same row, the second and thrid columns of Figure 4.1 are the ACF and PACF of the same series. Match series $r1$– $r3$ with series (a)–(c). Explain your answer briefly.

5

Examples in SPLUS

5.1 INTRODUCTION

Ideas from previous chapters are illustrated through two examples in this chapter. These examples are analyzed in detail using SPLUS programs. Some of the subtle points of SPLUS programs are also demonstrated in these examples.

5.2 EXAMPLE 1

This is an example of the yield of short-term government securities for 21 years for a country in Europe in a period in the 1950s and 1960s. The data are stored in the file `yields.dat` on the Web page for this book. Several observations can be made about this series.

```
>yield.ts_scan('yields.dat')
>ts.plot(yield.ts)
>acf(yield.ts)
>acf(yield.ts,30,type='partial')
```

There is an upward trend in the data, but there is not much seasonality. From Figure 5.1 we see that the data are clearly nonstationary. This phenomenon is quite common for financial series. Therefore, we should first difference the data as $W_t = (1-B)Y_t$. The plot of the differenced data and its corresponding ACF and PACF are given in Figure 5.2.

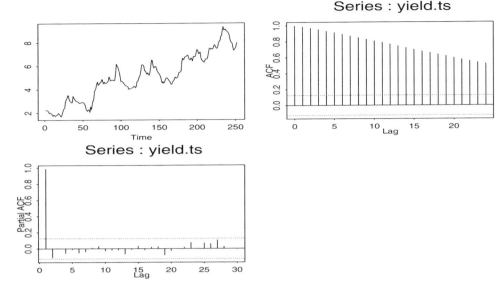

Fig. 5.1 Time series and ACF for the yield data.

```
>w_diff(yield.ts)
>ts.plot(w)
>acf(w)
>acf(w,30,type='partial')
```

Since there is a lag-one correlation in the ACF from Figure 5.2 that is different from zero, we may attempt an ARIMA(0,1,1) model for the original data, that is, fit an MA(1) for W_t as

$$(1 - B)Y_t = Z_t - \theta Z_{t-1}. \tag{5.1}$$

```
>w.1_arima.mle(w,model=list(order=c(0,0,1)))
>w.1$model$ma
[1] -0.4295783
>arima.diag(w.1)
```

This is also equivalent to fitting an ARIMA(0,1,1) to the demeaned series $Y_t - E(Y_t)$. One could have executed the command

```
w.2_arima.mle(yield.ts-mean(yield.ts),model=list(order=c(0,1,1)))
```

and obtain the same result. Note that Splus *reverses* the sign of θ and gives an estimate $\hat{\theta} = -0.43$. Therefore, the fitted equation becomes $(1 - B)\hat{Y}_t = Z_t + 0.43Z_{t-1}$. The last panel of Figure 5.2 gives the diagnostic plot of the fitted residuals. A simple ARIMA(0,1,1) model works reasonably well. This

EXAMPLE 1   *61*

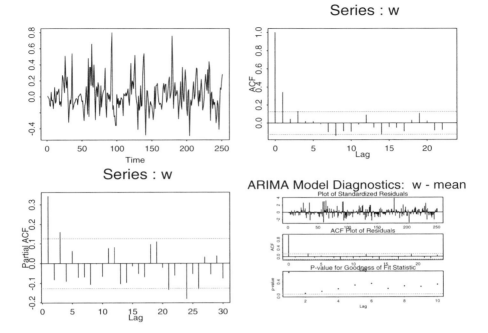

Fig. 5.2 Time series and ACF for the differenced yield data.

is a typical example for financial data, as most of them fit quite well with the random walk hypothesis (recall the discussion following Example 3.6). We can also perform simple forecasts with (5.1):

$$\hat{Y}_{n+1} = Y_n + \hat{Z}_{n+1} + 0.43\hat{Z}_n. \tag{5.2}$$

Since Z_{n+1} is unknown, we let $\hat{Z}_{n+1}$ as $E(Z_{n+1}) = 0$, so that

$$\hat{Y}_{n+1} = Y_n + 0.43\hat{Z}_n.$$

To complete the computation of $\hat{Y}_{n+1}$ from this equation, we need to specify $\hat{Z}_n$. This can be computed recursively as follows. Let the series $\{Y_1, \ldots, Y_n\}$ be given and let $Y_0 = Z_0 = 0$. Then according to (5.2),

$$\hat{Z}_1 = Y_1 - Y_0 - 0.43\hat{Z}_0 = Y_1,$$
$$\hat{Z}_2 = Y_2 - Y_1 - 0.43\hat{Z}_1,$$
$$\hat{Z}_3 = Y_3 - Y_2 - 0.43\hat{Z}_2,$$
$$\vdots$$
$$\hat{Z}_n = Y_n - Y_{n-1} - 0.43\hat{Z}_{n-1}.$$

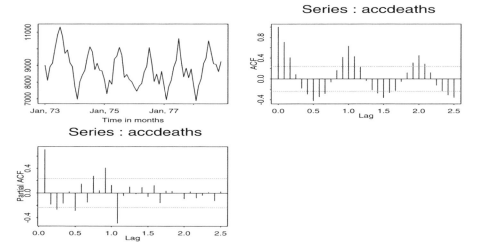

Fig. 5.3   Time series plots, ACF, and PACF of `accdeaths`.

After obtaining $\hat{Z}_n$ in this manner, we can subsitute this value into (5.2) to obtain $\hat{Y}_{n+1}$.

5.3   EXAMPLE 2

This is an example concerning the number of accidental deaths on U.S. highways between 1973 and 1978. The data are stored in the file `accdeaths.dat` on the Web page for this book. Here is an illustration of the SPLUS commands that can be used to analyze this series:

```
>accdeaths_scan('accdeaths.dat')
>ts.plot(accdeaths)
>acf (accdeaths,30)
>acf(accdeaths,30,type='partial')
```

By inspecting the ACF in Figure 5.3, a clear seasonal pattern is detected at lags 12, 24, 36, and so on. This is kind of intuitive, due to the high traffic during summer months. Note that we have a choice here; we can either do a seasonal smoothing as discussed in Chapter 1, or perform a seasonal differencing to account for the yearly effect. We choose the latter approach and perform a seasonal differencing at 12 lags for the data.

```
>dacc <- diff(accdeaths,12)
>ts.plot(dacc)
>acf(dacc)
>acf(dacc,30,type='partial')
```

EXAMPLE 2   63

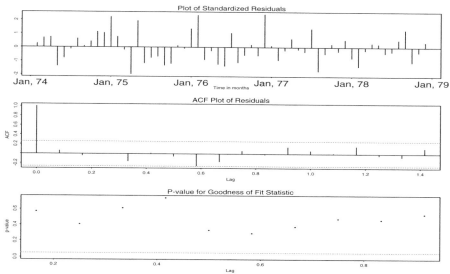

Fig. 5.4   Time series, ACF, and PACF of differenced data.

By inspecting the ACF of dacc in Figure 5.4, significant correlations up to lag 6 are detected. This leads to further differencing.

```
>ddacc <- diff(dacc)
>ts.plot (ddacc)
>acf(ddacc)
>acf(ddacc,30,type='partial')
```

Significant values of the ACF are detected at lags 1 and 12 in Figure 5.5 for ddacc, indicating a possible MA(1) × MA(12) model. Now entertain a SARIMA model with this preliminary information. We begin by fitting a SARIMA$(0, 0, 1) \times (0, 0, 1)_{12}$ model to the demeaned series.

```
>ddacc.1 <- arima.mle (ddacc-mean (ddacc), model = list
+(list(order = c (0, 0, 1)), list(order = c(0, 0, 1),
+ period =12)))
>ddacc.1 $model[[1]]$ ma
> 0.4884   (This is the MA parameter for the MA(1) part)
>ddacc.1 $model[[2]]$ ma
> 0.5853 (This is the MA(12) parameter for the MA(12) part)
>ddacc.1$aic
> 852.73
```

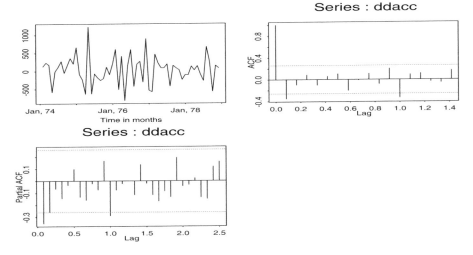

Fig. 5.5   Time series, ACF, and PACF of twice differenced data.

Essentially, what we have done is

$$W_t = (1 - B)(1 - B^{12})X_t,$$
$$V_t = W_t - \bar{W}, \quad \bar{W} = 28.8,$$
$$V_t = (1 - \theta B)(1 - \Theta B^{12})Z_t,$$
$$W_t = 28.83 + (1 - 0.488B)(1 - 0.5853B^{12})Z_t.$$

In other words, the model attempted so far is

$$(1 - B)(1 - B^{12})X_t = 28.83 + (1 - 0.488B)(1 - 0.5853B^{12})Z_t,$$

with an AIC value 852.73. If we want to get the standardized residuals, ACF plots, and p-values of Portmanteau statistics for various lags, use the command

```
>arima.diag(ddacc.1)
```

to generate Figure 5.6. Alternatively, try to fit an MA(13) to $W_t - \bar{W}$, and we should expect a large number of coefficients to be zeros. To do this, we proceed as follows:

```
>ddacc.2 <- arima.mle(ddacc-mean(ddacc),model=list
+(order=c(0,0,13)))
>ddacc.2$model$ma
```

This command gives the estimated values of all 13 MA parameters:

EXAMPLE 2   65

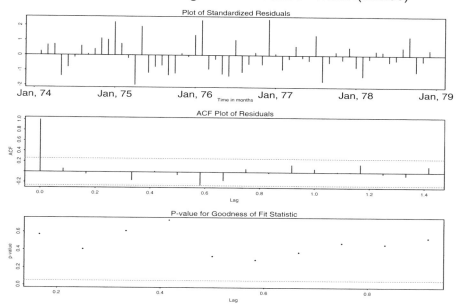

Fig. 5.6 Diagnostics and Portmanteau statistics for ddacc.1.

0.5	-0.025	0.067	0.085	-0.13	**0.19**
0.16	0.012	-0.11	-0.03	-0.09	**0.611**
-0.4					

```
>a <- ddacc.2$model$ma/(2*sqrt(diag(ddacc.2$var.coef)))
```

The object a consists of the two-standard-deviation ratios $\hat{\theta}_j/2\hat{\sigma}_j$ for $j = 1,\ldots,13$. Hence, if a particular entry of a is bigger than 1 in magnitude, it would indicate that the particular coefficient is significantly different from zero. The values of a are given as follows:

2.17	-0.11	0.30	0.38	-0.56	**0.83**
0.73	0.05	-0.50	-0.12	-0.43	**2.73**
-1.65					

Here, we observe $\theta_1, \theta_{12}, \theta_{13}$, and possibly θ_6 to be significantly different from zero. Refit the data by specifying $\theta_1, \theta_{12}, \theta_{13}$ to be the only nonzero values.

```
>ddacc.3 <- arima.mle(ddacc-mean(ddacc),model=list
+ (order=c(0,0,13),ma.opt=c(T,F,F,F,F,F,F,F,F,F,F,T,T)))
```

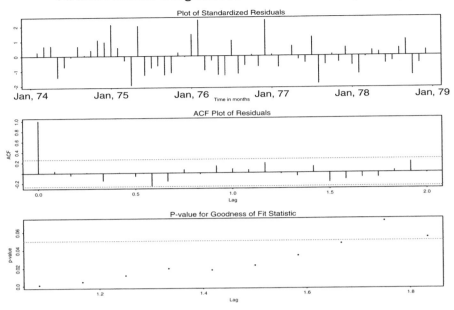

Fig. 5.7 Diagnostics and Portmanteau statistics for `ddacc.3`.

```
>ddacc.3$model$ma
>arima.diag(ddacc.3)
```

The estimated values are $\theta_1 = 0.46$, $\theta_{12} = 0.63$, $\theta_{13} = -0.22$, with $q = 3$. The AIC value is AIC $= 854.47 = -2\log L(\hat{\sigma}^2) + 2(p+q+1)$, where $p = 0$ and $q = 13$. Therefore, $-2\log L(\hat{\sigma}^2) = 854.47 - 28$, which implies that the correct AIC should be equal to $854.47 - 28 + 8 = 834.47$ since there are only three independent parameters and the correct $q = 3$. The p-values of Portmanteau statistics for this particular model seem to be quite poor (see Figure 5.7). As a further attempt, we try a model by including a possible nonzero θ_6.

```
>ddacc.4 <- arima.mle(ddacc-mean(ddacc),model=list(order=
+c(0,0,13), ma.opt=C(T,F,F,F,F,T,F,F,F,F,F,T,T)))
>ddacc.4$model$ma
> 0.60 ... 0.41 ... 0.67, -0.47
> AIC = 869.85
```

This time, $q = 4$ instead of 3, giving a revised AIC value as $869.85 - 18 = 851.85$. The diagnostic statistics in Figure 5.8 becomes more reasonable this time. The overall model is then

$$(1-B)(1-B^{12})X_t = 28.83 + (1 - 0.608B - 0.411B^6 - 0.677B^{12} + 0.473B^{13})Z_t.$$

EXAMPLE 2   67

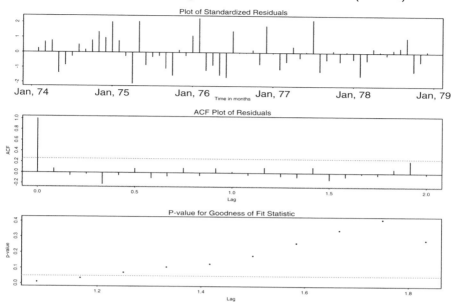

ARIMA Model Diagnostics: ddacc - mean(ddacc)

Fig. 5.8 Diagnostics and Portmanteau statistics for `ddacc.4`.

To perform a forecast, one can use SPLUS with the preceding fitted model as follows:

```
>ddacc.fore <- arima.forecast(ddacc-mean(ddacc),
+ n=6,model=ddacc.4$model)
>ddacc.fore$mean <- ddacc.fore$mean + mean(ddacc)
>67.771946 218.134433 -163.567978 9.058976 108.256377 62.774339
```

The preceding command forecasts the mean values of the series $W_t + \bar{W}$ for $t = 73, \dots , 78$. However, since we are interested in the values of X_t, we have to *undifference* the forecasted values $\hat{W}_t$ to get the values of X_t back. Thus, for $t = 73, \dots , 78$,

$$\hat{X}_t = \hat{W}_t + \hat{X}_{t-1} + \hat{X}_{t-12} - \hat{X}_{t-13}.$$

This gives the forecasted values listed in Table 5.1.

Table 5.1 Forecasted and Actual Values of $X_t, t = 73, \dots , 78$

t	73	74	75	76	77	78
Forecasted values $\hat{X}_t$	8348	7622	8357	8767	9798	10,180
Observed values X_t	7798	7406	8363	8460	9217	9316

5.4 EXERCISES

1. Reload the airline sales data in Exercise 3 of Chapter 1, which are listed on the Web page for this book under the file `airline.dat`.

   (a) Perform a time series analysis on the airline data set.

   (b) Forecast the next 12 values from the end of the data set with corresponding 95% forecasting intervals. You should also plot out the graph of the forecasts.

2. Perform a time series analysis on the weekly exchange rate between the U.S. dollar and the pound sterling for the years 1980–1988. This data set can be found on the Web page for this book under the file `exchange.dat`. Use the examples presented in this chapter as outlines to conduct your analysis.

6

Forecasting

6.1 INTRODUCTION

Having observed a time series $\{Y_1, \ldots, Y_n\}$, we are usually interested in forecasting a future value Y_{n+h}. It will be useful to distinguish between two types of forecast, the ex post and ex ante forecasts. The *ex post forecasts* observations when the "future" observations are known for certain during the forecasting period. It is used as a means to check against known data so that the forecasting model can be evaluated. On the other hand, the *ex ante forecasts* observations beyond the present, and in this case, the future observations are not available for checking. Suppose that we observe $\{Y_1, \ldots, Y_n\}$; we may use $\{Y_1, \ldots, Y_T\}$ $(T < n)$ to estimate a model and use the estimated model to forecast $Y_{T+1}, \ldots, Y_n$. These are ex post forecasts since we can use them to compare against the observed $Y_{T+1}, \ldots, Y_n$. The estimation period in this case is T. On the other hand, when we forecast $Y_{n+1}, \ldots, Y_{n+h}$ for $h > 0$, we are doing ex ante forecasts. After fitting a model, we estimate a future value Y_{n+h} at time n by $\hat{Y}_n(h)$ based on the fitted model, while the actual value of Y_{n+h} is unknown. The forecast is h steps ahead; h is known as the *lead time* or *horizon*. In practice, there are different ways to perform forecasts and we discuss several of these methods in the following sections. We begin by introducing three important quantities. The first is the forecast function $\hat{Y}_n(h)$, which is also denoted by $\hat{Y}(n, h), \hat{Y}_{n+h}$ or Y_{n+h}^n. Formally, it is defined as

$$\hat{Y}_{n+h} = E(Y_{n+h}|Y_n, \ldots)$$
$$= P_{\overline{sp}\{Y_n, \ldots\}}Y_{n+h}$$
$$= P_n Y_{n+h}$$
$$= Y_{n+h}^n. \tag{6.1}$$

Note that this is the conditional mean of Y_{n+h} given the history. It can be shown that this is the best estimator in terms of the mean square error (see Exercise 1). The second quantity is the corresponding forecast error, which is defined as

$$e_n(h) = Y_{n+h} - Y_{n+h}^n, \tag{6.2}$$

while the third is the variance of this error given by

$$P_{n+h}^n = E[(Y_{n+h} - Y_{n+h}^n)^2|Y_n, \ldots]. \tag{6.3}$$

Again, this variance is simply the conditional variance of the forecast error based on the history of the data.

6.2   SIMPLE FORECASTS

If the underlying time series has a simple trend such as those discussed in Chapter 1, Y_n can be forecasted by means of simple extrapolation. This is most useful when long-term forecasting is desired, where it is unlikely to be worthwhile to fit a complicated model. Specifically, consider $Y_n = m_n + X_n$, the signal plus noise model. Suppose m_n is constant, so that the series is stationary. Then

$$Y_{n+h}^n = P_n Y_{n+h} = \hat{m}_{n+h} = \hat{m}_n, \tag{6.4}$$

where the last quantity, $\hat{m}_n$, can be estimated by exponential smoothing. To conduct exponential smoothing, recall from Chapter 1 that the estimated trend at time n is expressed as a convex combination of the current observation and the previous estimate as

$$\hat{m}_{t+1} = \alpha Y_{t+1} + (1-\alpha)\hat{Y}_{t+1}$$
$$= \alpha Y_{t+1} + (1-\alpha)\hat{m}_{t+1}$$
$$= \alpha Y_{t+1} + (1-\alpha)\hat{m}_t \quad [\text{using } (6.4)], \quad t = 1, \ldots, n,$$

with $\hat{m}_1 = Y_1$. By iterating this recursion, it can be seen easily that the current estimated value m_n is simply a weighted average of the data:

$$\hat{m}_t = \sum_{j=0}^{t-2} \alpha(1-\alpha)^j Y_{t-j} + (1-\alpha)^{t-1}Y_1.$$

When the trend is not constant, one simple generalization of the forecast function is

$$P_nY_{n+h} = \hat{a}_n + \hat{b}_nh,$$

where a_n and b_n designate the level and the slope of the trend function at time n. A simple generalization of the exponential smoothing, known as the *Holt–Winters algorithm*, can be derived to account for such a situation. To illustrate this idea, first forecast the level a_{n+1} by exponentially smoothing the current value and the last forecasted value as

$$\hat{a}_{t+1} = \alpha Y_{t+1} + (1-\alpha)\hat{Y}_{t+1}$$

$$= \alpha Y_{t+1} + (1-\alpha)(\hat{a}_t + \hat{b}_t) \text{ for } t = 2,\ldots,n.$$

For the slope, first consider

$$Y^n_{n+2} = P_nY_{n+2} = \hat{a}_n + \hat{b}_n 2.$$

On the other hand, when Y_{n+1} is available, we can compute

$$Y^{n+1}_{n+2} = P_{n+1}Y_{n+2} = \hat{a}_{n+1} + \hat{b}_{n+1}.$$

Since both Y^n_{n+2} and Y^{n+1}_{n+2} are estimating the same future value Y_{n+2}, it is reasonable to expect them to be close. By equating these two quantities, we see that $\hat{b}_{n+1} \sim \hat{a}_n - \hat{a}_{n+1} + 2\hat{b}_n$. This idea motivates writing $\hat{b}_{t+1}$ as a convex combination of $\hat{a}_{t+1} - \hat{a}_t$ and $\hat{b}_t$ as

$$\hat{b}_{t+1} = \beta(\hat{a}_{t+1} - \hat{a}_t) + (1-\beta)\hat{b}_t \quad \text{for } t = 2,\ldots,n.$$

By combining the recursions for the estimated levels and slopes and fixing the initial conditions $\hat{a}_2 = Y_2$ and $\hat{b}_2 = Y_2 - Y_1$, $\hat{a}_n$ and $\hat{b}_n$ can be computed so that forecast of Y^n_{n+h} can be obtained. Similar to exponential smoothing, the values α and β are chosen such that the sum of squares of the prediction error $\sum_{i=2}^{n} e_i^2$ is minimized. In practice, α and β are found to be lying between 0.1 and 0.3. Further generalization of the Holt and Winters method for seasonal trends can be found in Kendall and Ord (1990).

6.3 BOX AND JENKINS APPROACH

This procedure amounts to fitting an ARIMA model and using it for forecasting purposes. It is discussed in detail in Abraham and Ledolter (1983). Let Y_t follow a causal and invertible ARMA(p,q) model $\phi(B)Y_t = \theta(B)Z_t$. Then $Y_t = \sum_{i=0}^{\infty}\psi_iZ_{t-i}$, $\psi_0 = 1$. Consider the following conditional expectations:

$$E(Y_s|Y_n,\ldots) = \begin{cases} Y_s, & s \le n, \\ Y^n_s, & s > n. \end{cases}$$

$$E(Z_s|Y_n,\ldots) = \begin{cases} Z_s, & s \le n, \\ 0, & s > n. \end{cases}$$

The last equation follows from the fact that $Y_n = \sum_{i=0}^{\infty} \psi_i Z_{n-i}$, so that $E(Z_s|Y_n, \dots) = E(Z_s|Z_n, \dots)$. Furthermore, since the forecast function satisfies

$$Y_{n+h}^n = E(Y_{n+h}|Y_n, \dots) = \sum_{i=0}^{\infty} \psi_i E(Z_{n+h-i}|Y_n, \dots) = \sum_{i=h}^{\infty} \psi_i Z_{n+h-i},$$

it follows that the forecast error becomes

$$e_n(h) = Y_{n+h} - Y_{n+h}^n = \sum_{i=0}^{h-1} \psi_i Z_{n+h-i}.$$

In particular, the forecast error variance is given by

$$P_{n+h}^n = E(e_n(h)^2|Y_n, \dots) = \sigma_Z^2 \sum_{i=0}^{h-1} \psi_i^2.$$

Therefore, as long as the forecast errors (innovations) are normally distributed, a $(1 - \alpha)$ probability interval for the forecast values can be constructed as

$$Y_{n+h}^n \pm z_{\alpha/2}\sqrt{P_{n+h}^n},$$

where z_α denotes the αth percentile of a standard normal random variable.

Example 6.1 *Suppose that the following* SARIMA$(1, 0, 0) \times (0, 1, 1)_{12}$ *has been fitted to a data set:*

$$Y_n = Y_{n-12} + \phi(Y_{n-1} - Y_{n-13}) + Z_n - \theta Z_{n-12}. \tag{6.5}$$

Since $E(Z_{n+1}|Y_n, \dots) = 0$, we have

$$\begin{aligned} Y_{n+1}^n &= E(Y_{n+1} \mid Y_n, Y_{n-1}, \dots) \\ &= Y_{n-11} + \phi(Y_n - Y_{n-12}) - \theta Z_{n-11}. \end{aligned}$$

Therefore,

$$\begin{aligned} Y_{n+2}^n &= E(Y_{n+2}|Y_n, Y_{n-1}, \dots) \\ &= E(Y_{n-10} + \phi(Y_{n+1} - Y_{n-11}) + Z_{n+2} - \theta Z_{n-10}|Y_n, \dots) \\ &= Y_{n-10} + \phi(Y_{n+1}^n - Y_{n-11}) - \theta Z_{n-10}. \end{aligned} \tag{6.6}$$

This procedure can be continued recursively. For example, if Y_{n+1} is available, by virtue of (6.6),

$$\begin{aligned} Y_{n+2}^{n+1} &= Y_{n-10} + \phi(Y_{n+1} - Y_{n-11}) - \theta Z_{n-10} \\ &= Y_{n-10} + \phi(Y_{n+1}^n - Y_{n-11}) - \theta Z_{n-10} - \phi Y_{n+1}^n + \phi Y_{n+1} \\ &= Y_{n+2}^n + \phi(Y_{n+1} - Y_{n+1}^n) \\ &= Y_{n+2}^n + \phi Z_{n+1}. \end{aligned}$$

$\square$

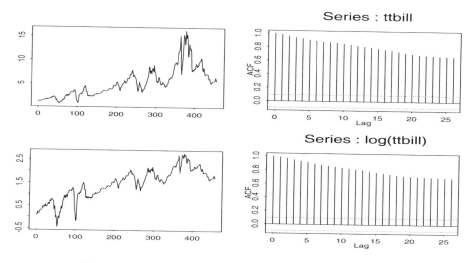

Fig. 6.1   Time series plots, ACF of T-bills, and log(T-bills).

For a general ARMA(p, q) model, similar equations can be used to calculate the values of the forecast. Again, approximations for the values of Z_n appearing in the forecasts are computed recursively for $n = p+1, p+2, \ldots$ by solving for Z_n in the ARMA equation and by making the assumptions that $Z_n = 0$ for $n \leq p$.

6.4   TREASURY BILL EXAMPLE

Consider the series that consists of the monthly interest rate on three-month government Treasury bills from the beginning of 1950 through June 1988. The Web site at the Federal Reserve Bank at St. Louis,

http://www.stls.frb.org/fred/data/business.html

contains many interesting long-term bond yield data. This data set is stored under the file ustbill.dat on the Web page for this book. The time series plot, the ACF, and the log-transformed series are given in Figure 6.1.

It is clear from the plots that the data are nonstationary in both mean and variance. As a first attempt, consider the difference in the logs of the data; call this series "dlntbill" and designate it as X_t. From Figure 6.2, both ACF and PACF show that there are strong correlations at lags 1, 6, and perhaps 17. Since the lag 17 correlation is only marginally significant, which may be caused by a number of factors (nonstationary variance, for example), we start with an AR(6) model, then an MA(6) model, and finally, an ARMA(6,6) model for dlntbill. In terms of the diagnostics and the AIC values, both ARMA(6,6)

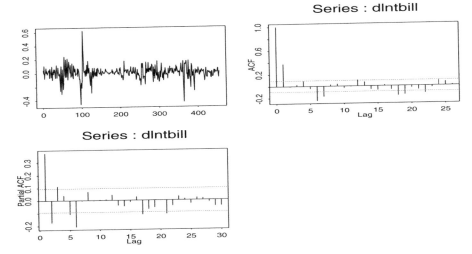

Fig. 6.2 Time series plots, ACF, and PACF of X_t.

and MA(6) compare less favorably than the AR(6). The diagnostic plots of the residuals of the fitted AR(6) model are given in Figure 6.3.

Consequently, we choose the AR(6) model as the fitted model and attempt to forecast the future values based on it. In summary, the fitted model is

$$\phi(B)Y_t = Z_t,$$

where $Y_t = W_t - \bar{W}, W_t = (1 - B)X_t$, $X_t = \log(\text{tbill})$ at month t, and $\phi(B) = 1 + 0.45B - 0.2B^2 + 0.09B^3 + 0.04B^4 - 0.01B^5 - 0.21B^6$. The SPLUS program for this analysis follows.

```
> tsplot(ttbill)
> acf(ttbill)
> tsplot(log(ttbill))
> acf(log(ttbill))
> dlntbill_diff(log(ttbill))
> tsplot(dlntbill)
> acf(dlntbill)
> acf(dlntbill,30,'partial')
> d3_arima.mle(dlntbill-mean(dlntbill),model=list
+ (order=c(6,0,0)))
> d3$model$ar
[1]   0.44676167 -0.19584331   0.09226429   0.04264963 -0.01214136
-0.20785506
> d3$aic
[1] -1027.108
> arima.diag(d3)
```

ARIMA Model Diagnostics: dlntbill - mean(dlntbill)

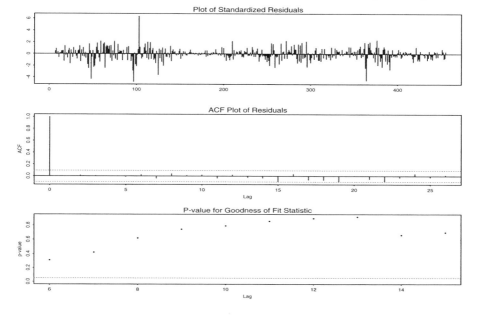

Fig. 6.3 Diagnostic plots of the fitted residuals from the AR(6) model.

```
> d3fore_arima.forecast(dlntbill-mean(dlntbill),n=6,
+ model=d3$model)
> d3fore
$mean:
[1]  0.017246377 -0.015810111 -0.023524633  0.005293435
     0.022376072
[6]  0.003773632
> d33_exp(d3fore$mean + mean(dlntbill))
> tfore_c(0,0,0,0,0,0)
> tfore[1]_ttbill[456]*d33[1]
> for (i in 2:6){
+ tfore[i]_tfore[i-1]*d33[i]}
> tfore
[1] 5.892155 5.821250 5.707001 5.758577 5.910733 5.955093
> tsplot(ttbill[457:462],tfore)
> leg.names_c('Actual','Forecast')
> legend(locator(1),leg.names,lty=c(1,2))
```

The forecasted and observed values for the first six months of 1988 are given in Table 6.1. Note that the SPLUS program accomplishes the following steps:

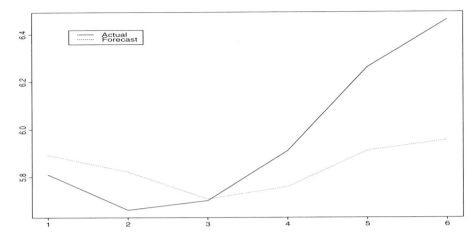

Fig. 6.4   Plots of actual and forecasted values.

1. Fit an AR(6) to the series $W_t - \overline{W}$ and forecast the next six values from this model as $\hat{Y}_{t+h} = \hat{W}_{t+h} - \overline{W}$. Therefore, $\hat{W}_{t+h} = \hat{Y}_{t+h} + \overline{W}$ for $h = 1, \ldots, 6$.

2. Recover X_t from W_t as $\hat{X}_{t+h} = \hat{W}_{t+h} + \hat{X}_{t+h-1} = \hat{Y}_{t+h} + \overline{W} + \hat{X}_{t+h-1}$.

3. Recover $\hat{Tbill}_{t+h} = e^{\hat{X}_{t+h}} = \exp(\hat{Y}_{t+h} + \overline{W} + \hat{X}_{t+h-1})$.

As can be seen from Figure 6.4, the model forecasts the first three months in 1988 reasonably well. However, when interest rates started hiking in April 1988, the fitted model did not have enough predictability to pick up this sharp rise. This is one of the shortcomings of using an ARIMA model to forecast a volatile financial series. Nevertheless, the overall trend of the forecasted values still follows the series observed. We have only modeled the data set by means of its own internal dynamics without taking account of any exogeneous information in this example. There are other ways to model fixed-income assets, commonly known as *term-structure models*, which have been studied extensively in the finance literature. Further details can be found in Chapter 10 of Campbell, Lo, and MacKinlay (1997).

Table 6.1   Forecasted and Actual Rates of T-Bills in 1988

t	Jan.	Feb.	Mar.	Apr.	May	June
Forecasted rates	5.89	5.82	5.71	5.76	5.91	5.96
Observed rates	5.81	5.66	5.70	5.91	6.26	6.46

6.5   RECURSIONS*

Suppose that a zero mean stationary time series $\{Y_1, \ldots, Y_n\}$ is given. In this section we mention two commonly used algorithms for calculating the prediction of Y_{n+1}. Let $\{Y_1, \ldots, Y_n\}$ be a zero mean stationary time series. The first algorithm makes use of the Durbin–Levinson recursion and expresses the one-step-ahead prediction of Y_{n+1} as the data, so that

$$\hat{Y}_{n+1} = Y_{n+1}^n = P_n Y_{n+1} = \sum_{j=1}^{n} \phi_{nj} Y_{n+1-j}, \tag{6.7}$$

where the coefficients ϕ_{nj} are determined via the Durbin–Levinson recursion, and the last coefficient, ϕ_{nn}, is the PACF of the series. Note that this expression is to write the predictor in terms of a linear combination of the data $Y_1, \ldots, Y_n$.

The second algorithm expresses $\hat{Y}_{n+1}$ as a linear combination of the one-step-ahead prediction error (innovation) $Y_j - \hat{Y}_j$, $j = 1, \ldots, n$, as

$$\hat{Y}_{n+1} = \sum_{j=0}^{n} \theta_{nj}(Y_{n+1-j} - \hat{Y}_{n+1-j}), \tag{6.8}$$

where the coefficients θ_{nj} can be computed recursively by means of the innovation algorithm. This algorithm is particularly useful when we have an ARMA(p, q) model so that θ_{nj} are functions of the underlying parameters $\phi, \boldsymbol{\theta}$. It is used to establish an on-line recursion to evaluate the likelihood function for an ARMA(p, q) model. Further discussions about these methods can be found in Chapter 5 of Brockwell and Davis (1991).

6.6   EXERCISES

1. Suppose we wish to find a prediction function $g(X)$ that minimizes

   $$\text{MSE} = E[(Y - g(X)^2],$$

   where X and Y are jointly distributed random variables with a joint density $f(x, y)$. Show that the MSE is minimized by the choice

   $$g(X) = E(Y|X).$$

2. Let X_t be an ARMA(1,1) model

   $$(1 - \phi B)(X_t - \mu) = Z_t - \theta Z_{t-1}.$$

---

*Throughout this book, an asterisk indicates a technical section that may be browsed casually without interrupting the flow of ideas.

(a) Derive a formula for forecast function X_{n+h}^n in terms of (ϕ, θ) and Z_t.

(b) What would X_{n+h}^n tend to as h tends to ∞? Can you interpret this limit?

(c) Find the formula for $\text{var}[e_t(h)]$. Again, interpret this formula as h tends to ∞.

3. Write a SPLUS program to perform a Holt–Winters prediction of the accidental deaths series (ignoring the seasonal component) discussed in Section 5.3 for $t = 73, \dots, 78$. Compare your results with those given in Table 5.1 and comment on the differences.

4. Without a logarithm, fit an ARIMA model to the original monthly Treasury rates in Section 6.4 and compare your forecasted values with the model given in the example.

5. Let $\{Y_t\}$ be a stationary time series with mean μ. Show that

$$P_{\overline{sp}\{1, Y_1, \dots, Y_n\}} Y_{n+h} = \mu + P_{\overline{sp}\{X_1, \dots, X_n\}} X_{n+h},$$

where $X_t = Y_t - \mu$.

7

Spectral Analysis

7.1 INTRODUCTION

Although spectral analysis has many important applications in physical sciences and engineering, its applications in the field of finance are still limited. Nevertheless, given its historical background and applications in other disciplines, a basic understanding about its nature will be fruitful. In this chapter we discuss briefly the idea of spectral analysis in time series. Further details about spectral analysis can be found in the books by Bloomfield (2000), Koopmans (1995), and Priestley (1981).

7.2 SPECTRAL REPRESENTATION THEOREMS

Most time series exhibit some periodic patterns that can roughly be represented by a sum of harmonics. To understand this type of approximation, we introduce a few terms. The first one specifies the rate at which a series oscillates in terms of cycles, where a *cycle* is defined as a sine or cosine wave defined over a time interval of extent 2π. The oscillation rate ω is the *frequency*, which is defined as cycles per unit time. Sometimes, it is more convenient to define frequency as radians per unit time λ [1 radian $= 1/(2\pi)$ of a cycle] and we have the relation

$$\lambda = 2\pi\omega.$$

Finally, the *period* T_0 of an oscillating series is defined as the length of time required for one full cycle and can be expressed as the reciprocal of the frequency,

$$T_0 = 1/\lambda.$$

As an example, let $B(\lambda_j)$ be a sequence of independent random variables defined over a given sequence of frequencies $-\pi < \lambda_1 < \cdots < \lambda_n \leq \pi$ in radians per unit time. Consider the process defined by

$$Y_t = \sum_{j=1}^{n} B(2\pi\omega_j) \, \sin(2\pi\omega_j t + \phi)$$

$$= \sum_{j=1}^{n} B(\lambda_j) \, \sin(\lambda_j t + \phi)$$

$$\cong \sum_{j=1}^{n} A(\lambda_j) e^{it\lambda_j},$$

where the last approximation follows from Euler's identity of complex numbers. The quantities $A(\lambda_j)$, $j = 1, \ldots, n$, are random amplitudes that are assumed to be independent random variables with $E(A(\lambda_j)) = 0$, var $(A(\lambda_j)) = \sigma_j^2$, $j = 1, \ldots, n$, and ϕ is a fixed phase shift among these harmonics. Such a process $\{Y_t\}$ is stationary. To check this, first observe that $EY_t = 0$. Further, if we denote $\bar{Y}_t$ as the complex conjugate of Y_t, the covariance function of Y_t is given by

$$\gamma(k) = EY_{t+k}\bar{Y}_t$$

$$= E\left(\sum_{j=1}^{n} A(\lambda_j) e^{i(t+k)\lambda_j} \sum_{l=1}^{n} \bar{A}(\lambda_l) e^{-it\lambda_l} \right)$$

$$= \sum_{j=1}^{n} \sigma_j^2 e^{-ik\lambda_j}.$$

Hence, the process $\{Y_t\}$ is stationary.

Define a function $F(\lambda) = \sum_{j:\lambda_j \leq \lambda} \sigma_j^2$ for $-\pi < \lambda \leq \pi$. Then F is a nondecreasing function and we can express $\gamma(k)$ as a Riemann–Stieltjes integral of F as

$$\gamma(k) = \int_{-\pi}^{\pi} e^{ik\lambda} \, dF(\lambda). \tag{7.1}$$

In particular, $F(\pi) = \sum_{j=1}^{n} \sigma_j^2 = \text{var } Y_t$. F is called the *spectral distribution function* of $\{Y_t\}$. Furthermore, if F is absolutely continuous, the function $f(\lambda) = dF(\lambda)/d\lambda$ is known as the *spectral density function* of the $\{Y_t\}$. More generally, it can be shown that equation (7.1) holds for a wide class of stationary processes. The next three theorems, usually known as *spectral representation theorems*, are technical in nature. Since they are used primarily in deriving theoretical properties, we shall just state these theorems without pursuing the details. Interested readers may find further discussions in Priestley (1981).

The first theorem states that the covariance function of a stationary process is related to the spectral distribution function through a Fourier transform. This is sometimes known as the *spectral representation theorem* of the auto-covariance function.

Theorem 7.1 $\gamma(\cdot)$ *is the ACF of a stationary process* $\{Y_t\}$ *iff*

$$\gamma(k) = \int_{-\pi}^{\pi} e^{ik\lambda} \, dF(\lambda),$$

where F is a right continuous, nondecreasing function on $[-\pi, \pi]$ with $F(-\pi) = 0$. Further, if F is such that $F(\gamma) = \int_{-\pi}^{\gamma} f(\lambda) \, d\lambda$, then $f(\cdot)$ is called the spectral density function of $\{Y_t\}$.

The second theorem relates the process $\{Y_t\}$ itself to another stationary process $\{Z(\lambda)\}$ defined over the spectral domain. It is sometimes known as the spectral representation theorem of the process.

Theorem 7.2 *Every zero mean stationary process* $\{Y_t\}$ *can be written as*

$$Y_t = \int_{-\pi}^{\pi} e^{it\lambda} \, dZ(\lambda),$$

where $\{Z(\lambda)\}$ is a stationary process of independent increments.

The third theorem relates the spectral density function to the autocovariance function.

Theorem 7.3 *An absolutely summable $\gamma(\cdot)$ is the ACF of a stationary process $\{Y_t\}$ iff it is even and the function f is defined as*

$$f(\lambda) = \frac{1}{2\pi} \sum_{k=-\infty}^{\infty} \gamma(k) e^{-ik\lambda} \geq 0$$

for all $\lambda \in [-\pi, \pi]$, in which case f is called the spectrum or spectral density of $\gamma(\cdot)$.

Remark. It is easy to see that such an f satisfies the following properties:

1. f is even [i.e., $f(\lambda) = f(-\lambda)$].

2. $\gamma(k) = \int_{-\pi}^{\pi} e^{ik\lambda} f(\lambda) \, d\lambda = \int_{-\pi}^{\pi} \cos(k\lambda) f(\lambda) \, d\lambda.$

Example 7.1 *Let* $Y_t = \phi Y_{t-1} + Z_t$, $|\phi| < 1$, *an* AR(1) *process. Recall that* $\gamma(k) = \rho(k)/\gamma(0)$, *where* $\rho(k) = \phi^k$ *and* $\gamma(0) = \sigma^2/(1-\phi^2)$. *Now consider the function*

$$f(\lambda) = \frac{1}{2\pi} \sum_{k=-\infty}^{\infty} \gamma(k) e^{ik\lambda}.$$

In particular,

$$\frac{1}{\gamma(0)} f(\lambda) = \frac{1}{2\pi} \sum_{k=-\infty}^{\infty} \rho(k) e^{ik\lambda}$$

$$= \frac{1}{2\pi} \left(1 + \sum_{k=1}^{\infty} \phi^k e^{-ik\lambda} + \sum_{k=1}^{\infty} \phi^{-k} e^{ik\lambda} \right)$$

$$= \frac{1}{2\pi} \frac{1 - \phi^2}{1 - 2\phi \cos \lambda + \phi^2} \geq 0 \quad \textit{for all } \lambda \in (-\pi, \pi].$$

According to Theorem 7.3, $\gamma(k)$ *is the autocovariance function of a stationary process* $\{Y_t\}$, *which is the given* AR(1) *process.*   □

The graphs of f for different values of ϕ in this example are given in Figure 7.1.

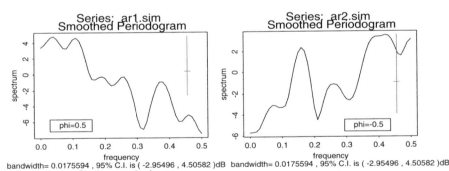

Fig. 7.1 Spectrum for an AR(1) model with different ϕ.

For a general $\mathrm{ARMA}(p, q)$ process, its spectral density function is given by

$$f(\lambda) = \frac{\sigma^2}{2\pi} \left| \frac{\theta(e^{-i\lambda})}{\phi(e^{-i\lambda})} \right|^2.$$

In particular, for the AR(1) case we have

$$f(\lambda) = \frac{\sigma^2}{2\pi} \frac{1}{|\phi(e^{-i\lambda})|^2},$$

where

$$|\phi(e^{-i\lambda})|^2 = (1 - \phi e^{i\lambda})(1 - \phi e^{-i\lambda}) = 1 - 2\phi \cos \lambda + \phi^2,$$

agreeing with Example 7.1. Since any "nice" function can be approximated by a rational function, the result above implies that for any given stationary process with a "nice" spectrum, it can be approximated by an ARMA process through spectral consideration. This is one of the reasons why an ARMA model is useful.

Example 7.2 *As a second example, let $\{Y_t\} \sim \mathrm{WN}(0, \sigma^2)$. Recall that $\rho(k) = 0$ for $k \neq 0$ and*

$$f(\lambda) = \frac{\sigma^2}{2\pi} \quad \forall \, \lambda \in (-\pi, \pi].$$

In other words, the spectrum f is flat for all frequencies, hence the term **white noise**. □

7.3 PERIODOGRAM

In practice, since $\rho(k)$ and $\gamma(k)$ are unknown, we must resort to estimation and construct

$$\hat{f}(\omega) = \frac{1}{2\pi} \sum_{k=-\infty}^{\infty} \hat{\gamma}(k) e^{ik\omega}.$$

However, evaluating $\hat{f}$ is difficult. Instead, consider the following function:

Definition 7.1

$$I(\omega) = \frac{1}{n} \left| \sum_{t=1}^{n} Y_t e^{-it\omega} \right|^2, \tag{7.2}$$

known as the **periodogram**.

Note that the periodogram is just a truncated expression of $\hat{f}$, as shown in the following result.

Theorem 7.4 *Let $\omega_j = 2\pi j/n$ be a nonzero frequency. Then*

$$I(\omega_j) = \sum_{|k|<n} \hat{\gamma}(k)e^{-ik\omega_j}.$$

Proof. According to Definition 7.1,

$$I(\omega_j) = \frac{1}{n}\sum_{t=1}^{n} Y_t e^{-it\omega_j} \sum_{t=1}^{n} \bar{Y}_t e^{it\omega_j}.$$

Recall the fact that $\sum_t e^{it\omega_j} = \sum_t e^{-it\omega_j} = 0$ if $\omega_j \neq 0$. Hence,

$$I(\omega_j) = \frac{1}{n}\sum_{t=1}^{n}\sum_{s=1}^{n}(Y_t - m)(\bar{Y}_s - \bar{m})\bar{e}^{i(s-t)\omega j}$$

$$= \sum_{|k|<n} \hat{\gamma}(k)e^{-ik\omega_j}. \qquad \square$$

To speed up the calculation of $I(\omega)$, we usually employ the fast Fourier transform algorithm. Again, details on this method can be found in Priestley (1981). Since the spectrum is estimated by $I(\omega)$, one immediately questions: Why would a periodogram be good? What property would $I(\omega)$ possess for a periodic series? Since a periodogram at a given Fourier frequency is the Fourier coefficient for the expansion of the data vector $Y = (Y_1, \ldots, Y_n)$ in terms of an orthonormal basis at that particular frequency, we expect the periodogram to have a strong contribution (level is high) when the data have a periodic component that is in sync with the given Fourier frequency. Specifically, consider the following example.

Example 7.3 *Let $Y_t = a\,\cos t\theta$; $\theta \in (0, \pi)$ be fixed and a, a random variable such that $E(a) = 0, \mathrm{var}(a) = 1$. For a fixed $\omega \in [-\pi, \pi]$, consider*

$$\sum Y_t \cos t\omega = \sum a \cos t\theta \cos t\omega$$

$$= \sum \frac{a}{2}[\cos t(\theta + \omega) + \cos t(\theta - \omega)].$$

Hence,

$$\sum Y_t \cos t\omega = \begin{cases} O(n) & \text{if } \theta = \omega, \\ O(1) & \text{if } \theta \neq \omega. \end{cases}$$

Equivalently,

$$I(\omega) = \begin{cases} O(n) & \text{if } \theta = \omega, \\ O(1) & \text{if } \theta \neq \omega. \end{cases} \qquad \square$$

In other words, for a given stationary series, use $I(\omega)$ to estimate the periodicity of the series. If the series is periodic, its periodogram attains a

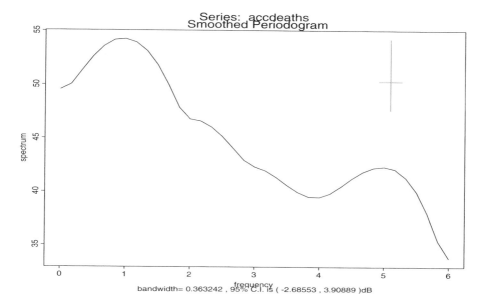

Fig. 7.2 Spectrum of the accidental death data set.

large value at $\omega = \theta$ and a small value otherwise. The periodogram of Example 5.2 reveals this feature in Figure 7.2.

```
>spectrum (accdeaths, spans=c(5,7))
```

This periodogram $I(\omega)$ has a peak at $\frac{1}{6}$ cycle. With a series of length $n = 72$, this translates to a period of 12. The scale of the periodogram in SPLUS is decibel, which is $10\log_{10}I(\omega)$. The command `spans=c(5,7)` is an input parameter from the user for smoothing the periodogram. One can adjust the shape of the periodogram by tuning this parameter. Details of this function can be found by typing `help(spectrum)`.

7.4 SMOOTHING OF PERIODOGRAM*

Since $I_n(\omega_j) = n^{-1}|\sum_{t=1}^{n} X_t e^{-it\omega_j}|^2$, $I_n(\omega_j)/2\pi$ would be an estimate of $f(\omega_j)$. We now extend the definition of I_n to the entire interval of $[-\pi, \pi]$.

Definition 7.2 *For any $\omega \in [-\pi, \pi]$, the periodogram is defined as*

$$I_n(\omega) = \begin{cases} I_n(\omega_k) & \text{if } \omega_k - \pi/n < \omega \le \omega_k + \pi/n \text{ and } 0 \le \omega \le \pi, \\ I_n(-\omega) & \text{if } \omega \in [-\pi, 0). \end{cases}$$

---

*Throughout this book, an asterisk indicates a technical section that may be browsed casually without interrupting the flow of ideas.

This definition shows that I_n is an even function that coincides with the original definition of periodogram at the nonnegative Fourier frequencies. For $\omega \in [0, \pi]$, let $g(n, \omega)$ be the multiple of $2\pi/n$ that is closest to ω, and let $g(n, \omega) = g(n, -\omega)$ for $-\pi \le \omega < 0$. Then Definition 7.2 can be written as

$$I_n(\omega) = I_n(g(n, \omega)) \quad \text{for all } \omega \in [-\pi, \pi]. \tag{7.3}$$

Asymptotic properties of I_n can be deduced and they can be found in Brockwell and Davis (1991). Since I_n is defined through some kind of interpolation, its graph would be spiky. For a better interpretation, we may want to smooth it. For example, consider the moving average version

$$\frac{1}{2\pi} \sum_{|k| \le m} \frac{1}{2m + 1} I_n(\omega_{j+k}).$$

This would be a smoothed estimate of $f(\omega_j)$ since it pools in the m nearby frequencies of ω_j. In general, we consider a smoothed version of I_n as

$$\hat{f}(\omega_j) = I_n(\omega) = \sum_{|k| \le m_n} W_n(k) I_n(\omega_{j+k}),$$

where $\{m_n\}$ is a sequence of positive integers and $\{W_n(\cdot)\}$ is a sequence of weight functions such that:

1. $m_n = o(n)$.

2. $W_n(k) = W_n(-k), \quad W_n(k) \ge 0$.

3. $\sum_{|k| \le m_n} W_n(k) = 1$ and $\sum_{|k| \le m_n} W_n^2(k) \to 0$.

Similar to the earlier discussion, in general, we can define a discrete spectral average estimator for any $\omega \in [-\pi, \pi]$ by

$$\hat{f}(\omega) = \hat{f}(g(n, \omega)) \tag{7.4}$$

$$= \frac{1}{2\pi} \sum_{|k| \le m_n} W_n(k) I_n(g(n, \omega) + 2k\pi/n), \tag{7.5}$$

where $g(n, \omega)$ is the multiple of $2\pi/n$ that is closest to ω. Here the sequence $\{m_n\}$ is known as the *bandwidth* (bin width) and the functions W_n are weight functions.

Under certain regularity conditions for a stationary linear process $Y_t = \sum_{j=0}^{\infty} \psi_j Z_{t-j}$, it can be shown that the smoothed estimate $\hat{f}$ has the following asymptotic properties:

1.

$$\lim_{n \to \infty} E(\hat{f}(\omega)) = f(\omega).$$

2.

$$\lim_{n\to\infty}\left(\sum_{|j|\leq m_n}W_n^2(j)\right)^{-1}\text{cov}(\hat{f}(\omega),\hat{f}(\lambda)) = \begin{cases} 2f^2(\omega) & \omega = \lambda = 0 \text{ or } \pi, \\ f^2(\omega) & 0 < \omega = \lambda < \pi, \\ 0 & \omega \neq \lambda. \end{cases}$$

This result indicates that there is a compromise between bias and variance of the spectral estimator. When m_n is large, the term $\sum W_n^2(k) \to 0$, which implies that the variance of $\hat{f}$ is small. The periodogram is smooth, but the bias may be large since $\hat{f}(\omega)$ depends on values of $I_n(\omega)$ at frequencies distant from ω. On the other hand, if m_n is small, the estimator will have a narrow frequency band and will give an estimator with small bias but with possibly large variance.

Alternatively, we can also define the lag window estimators as

$$\hat{f}_L(\omega) = \frac{1}{2\pi}\sum_{|h|\leq r}w(h/r)\hat{\gamma}(h)e^{-ih\omega},$$

where $w(x)$ is an even, piecewise continuous function of x satisfying

1. $w(0) = 1$.

2. $|w(x)| \leq 1$ for all x.

3. $w(x) = 0$ for $|x| > 1$.

The function w is called the *lag window* and $\hat{f}_L$ is called the *lag window spectral density estimator*. By taking the special case that $w(x) = 1$ and $r = n$, we have $\hat{f}_L(\omega_j) = (1/2\pi)I_n(\omega_j)$ for all nonzero Fourier frequencies ω_j.

Although the discrete spectral average estimator and the lag window estimator have different expressions, they can be linked by means of a spectral window, defined as

$$W(\omega) = \frac{1}{2\pi}\sum_{|h|\leq r}w(h/r)e^{-ih\omega}.$$

It can be shown that

$$\hat{f}_L(\omega) \sim \frac{1}{2\pi}\sum_{|j|\leq[n/2]}\frac{2\pi}{n}W(\omega_j)I_n(g(n,\omega)+\omega_j).$$

Consequently, it is a discrete spectral average estimator with weighting functions

$$W_n(j) = \frac{2\pi}{n}W(\omega_j).$$

It can further be shown that the lag window estimators have desirable asymptotic properties under certain regularity conditions; in particular:

1.

$$\lim_{n \to \infty} E(\hat{f}_L(\omega)) = f(\omega).$$

2.

$$\frac{n}{r} \text{var}(\hat{f}_L(\omega)) \to \begin{cases} 2f^2(\omega) \int_{-1}^{1} w^2(x) \, dx & \text{if } \omega = 0 \text{ or } \pi, \\ f^2(\omega) \int_{-1}^{1} w^2(x) \, dx & \text{if } 0 < \omega < \pi. \end{cases}$$

We shall now present a couple of examples for spectral windows.

Example 7.4 (Rectangular Window) *The rectangular window has the form*

$$w(x) = \begin{cases} 1 & \text{if } |x| \leq 1, \\ 0 & \text{otherwise.} \end{cases}$$

The corresponding spectral window is given by the Dirichlet kernel

$$W(\omega) = (2\pi)^{-1} \frac{\sin(r + \frac{1}{2}\omega)}{\sin(\omega/2)}.$$

In this case, the asymptotic variance is $(2r/n)f^2(\omega)$. ☐

Example 7.5 (Bartlett Triangular Window) *The Bartlett (triangular) window has the form*

$$w(x) = \begin{cases} 1 - |x| & \text{if } |x| \leq 1, \\ 0 & \text{otherwise.} \end{cases}$$

The corresponding spectral window is the Fejer kernel,

$$W(\omega) = (2\pi r)^{-1} \frac{\sin^2(r\omega/2)}{\sin^2(\omega/2)}.$$

The asymptotic variance in this case becomes $(2r/3n)f^2(\omega)$. ☐

Comparing these two examples, we see that the Bartlett window has a smaller asymptotic variance than the rectangular window. Other examples abound, including the Daniell window, Blackman–Tukey window, Parzen window, and so on. Each is an improvement on its predecessor in terms of asymptotic variance. Details can be found in Priestley (1981).

Finally, confidence bands for spectral estimators can be constructed by using the fact that $\nu \hat{f}(\omega_j)/f(\omega_j)$ is approximately distributed as a chi-squared distribution with ν degrees of freedom, where $\nu = 2/[\sum_{|k| \leq m_n} W_n^2(k)]$ is known as the *equivalent degrees of freedom* of the estimator. Consequently, for $0 < \omega_j < \pi$, an approximate 95% confidence interval is given by

$$\left(\frac{\nu \hat{f}(\omega_j)}{\chi^2_{0.975}}, \frac{\nu \hat{f}(\omega_j)}{\chi^2_{0.025}} \right).$$

7.5   CONCLUSIONS

Although spectral analysis is useful, it has limitations. First, spectral analysis deals with estimating the spectral density function, which lies in an infinite-dimensional function space from a time series of finite length. This is an example of the nonparametric density estimation problem. As such, we would not expect spectral estimation to deliver the same amount of accuracy as for a finite-dimensional parametric model: an ARMA, for example.

Second, it is important to realize that when inferences are drawn from the periodogram, we should pay more attention to its overall qualitative behavior rather than to microscopic details. After all, it is the general shape of the spectrum that matters most, uncovering periodicity, for example. In this context, applications of spectral analysis to econometric time series will be less useful than applications to physical sciences. Since the periodicity of most econometric time series, such as business cycles or quarterly effects, can easily be identified, spectral analysis in these areas will be of only marginal value. Uncovering unexpected periodicity in other areas would be more informative.

Third, prediction in the spectral domain has not been discussed in this chapter. This was once an active research area, and interested readers can find more discussion of this topic in Hannan (1970).

7.6   EXERCISES

1. Show that

$$\int_{-\pi}^{\pi} e^{i(k-h)\lambda}\,d\lambda = \begin{cases} 2\pi, & \text{if } k = h, \\ 0 & \text{otherwise.} \end{cases}$$

2. Let f be the spectral density function of a stationary process $\{X_t\}$ with absolutely summable autocovariance function $\gamma(\cdot)$ defined by

$$f(\lambda) = \frac{1}{2\pi}\sum_{h=-\infty}^{\infty} e^{-ih\lambda}\gamma(h), \quad -\pi < \lambda \le \pi.$$

   (a) Show that f is even [i.e., $f(\lambda) = f(-\lambda)$].

   (b) Consider the function

$$f_n(\lambda) = \frac{1}{2\pi n}E\left(\left|\sum_{r=1}^{n} X_r e^{-ir\lambda}\right|^2\right).$$

   Show that $f_n(\lambda) = (1/2\pi n)\sum_{|h|<n}(n-|h|)e^{-ih\lambda}\gamma(h)$ and that $f_n(\lambda)$ tends to $f(\lambda)$ as n tends to infinity.

   (c) Deduce from part (b) that $f(\lambda) \ge 0$.

(d) Show that

$$\gamma(k) = \int_{-\pi}^{\pi} e^{ik\lambda} f(\lambda)\ d\lambda = \int_{-\pi}^{\pi} \cos(k\lambda) f(\lambda)\ d\lambda.$$

3. (a) Let $\{X_t\}$ and $\{Y_t\}$ be two stationary processes with spectral density functions $f_X(\lambda)$ and $f_Y(\lambda)$, respectively. Show that the process $V_t = X_t + Y_t$ is also stationary with spectral density function

$$f_V(\lambda) = f_X(\lambda) + f_Y(\lambda).$$

(b) Let $V_t = X_t + Y_t$, where $\{X_t\}$ satisfies

$$X_t = \alpha X_{t-1} + W_t,$$

with $|\alpha| < 1$ and $\{Y_t\}$ and $\{W_t\}$ are independent white noise processes with zero mean and common variance σ^2. Find the spectral density function of $\{V_t\}$.

4. Compute and sketch the spectral density function of the process $\{X_t\}$ defined by

$$X_t = 0.99 X_{t-3} + Z_t, \quad Z_t \sim \text{WN}(0,1).$$

(a) Does the spectral density suggest that the sample path of $\{X_t\}$ will exhibit oscillatory behavior? If so, what is the approximate period of the oscillation?

(b) Simulate a realization of $\{X_1, \ldots, X_{60}\}$ and plot the realization. Does the plot support the previous conclusion? Also plot its periodogram.

(c) Compute the spectral density function of the process

$$Y_t = \frac{1}{3}(X_{t-1} + X_t + X_{t+1}).$$

Compare the numerical values of the spectral densities of $\{X_t\}$ and $\{Y_t\}$ at frequency $\omega = 2\pi/3$. What effect would you expect the filter to have on the oscillations of $\{X_t\}$?

(d) Apply the three-point moving average filter to the simulated realization. Plot the filtered series and the periodogram of the filtered series. Comment on your result.

5. Consider an AR(2) model

$$Y_t = \phi_1 Y_{t-1} + \phi_2 Y_{t-2} + Z_t, \quad Z_t \sim \text{WN}(0, \sigma^2).$$

(a) Derive an expression for the spectral density function of $\{Y_t\}$.

(b) Simulate a series of length 500 with $\phi_1 = 0.4$ and $\phi_2 = -0.7$ by using $Z_t \sim N(0, 1)$, i.i.d. Is such a series causal? Plot its periodogram and comment on its shape. Can you detect some periodicity? Why?

6. Let $\{X_t\}$ be the process defined by

$$X_t = A\cos(\pi t/3) + B\sin(\pi t/3) + Y_t,$$

where $Y_t = Z_t + 2.5Z_{t-1}$, $Z_t \sim WN(0, \sigma^2)$, A and B are uncorrelated random variables with mean zero and variance ν^2, and Z_t is uncorrelated with A and B for each t. Find the autocovariance function and the spectral distribution function of $\{X_t\}$.

7. Perform a spectral analysis on the Treasury bill data set discussed in Chapter 6. State all conclusions from your analysis and compare them with what you found earlier in a time-domain analysis.

8

Nonstationarity

8.1 INTRODUCTION

Different kinds of nonstationarity are often encountered in practice. Roughly speaking, a nonstationary time series may exhibit a systematic change in mean, variance, or both. We have developed some intuitive ideas regarding dealing with nonstationary time series. For instance, by inspecting the ACF, we may render a series to stationarity by differencing. Since the definition of nonstationarity varies, we focus our discussion on a special form of non-stationarity that occurs most often in econometrics and financial time series (i.e., nonstationarity in the mean level of the series). Before doing that, we start with a brief discussion of the transformation of nonstationary time se-ries with nonconstant variances. In the next chapter we study the notion of heteroskedasticity more systematically.

8.2 NONSTATIONARITY IN VARIANCE

Consider the situation where the mean level of the series is varied determin-istically, but the variance of the series is varying according to the mean level. Such a series can be expressed as

$$Y_t = \mu_t + Z_t,$$

where μ_t is a nonstochastic mean level but the variance of $\{Y_t\}$ has the form var $(Y_t) =$ var $(Z_t) = h^2(\mu_t)\sigma^2$ for some function h. Such an expression has the effect that the variance of $\{Y_t\}$ is proportional to its mean level μ_t. To account for such a phenomenon, we want to find a transformation g on $\{Y_t\}$ such the variance of $g(Y_t)$ is constant (i.e., to find a variance-stabilizing transformation). This can be accomplished by using a Taylor's approximation. Specifically, since

$$g(Y_t) \cong g(\mu_t) + (Y_t - \mu_t)g'(\mu_t),$$

we have

$$\text{var } (g(Y_t)) = [g'(\mu_t)]^2 \text{ var } (Y_t) = [g'(\mu_t)]^2 h^2(\mu_t)\sigma^2.$$

By specifying the variance of $g(Y_t)$ to be a fixed positive constant c (i.e., setting the right-hand side of this equation to a constant c), we obtain the relationship $g'(\mu_t) = 1/h(\mu_t)$.

As an example, if $h(\mu_t) = \mu_t$, then $g'(\mu_t) = 1/\mu_t$, which implies that $g(\mu_t) = \log(\mu_t)$, resulting in the usual logarithmic transformation. On the other hand, if $h(\mu_t) = (\mu_t)^{1/2}$, $g'(\mu_t) = (\mu_t)^{-1/2}$, and this implies that $g(\mu_t) = 2\mu_t^{1/2}$, resulting in the square-root transformation. In general, the Box–Cox transformation, $y^\lambda = (y^\lambda - 1)/\lambda$ if $\lambda \neq 0$ and $y^\lambda = \log(y)$ if $\lambda = 0$, can be used as an appropriate variance-stabilizing transformation. More details about the Box–Cox transformation can be found in Chapter 6 of Weisberg (1985).

8.3   NONSTATIONARITY IN MEAN: RANDOM WALK WITH DRIFT

When nonstationarity in the mean level is presented, the situation can be more complicated than nonstationarity in variance. Consider a linear trend plus a noise model

$$Y_t = \beta_0 + \beta_1 t + Z_t.$$

Differencing leads to

$$\Delta Y_t = \beta_1 + \Delta Z_t.$$

Although this model is stationary, it is no longer invertible. Another way to represent a change in mean level is to consider the following two models:

$$Y_t = \beta_0 + \beta_1 t + v_t \quad \text{(TS)},$$

$$Y_t = \beta_1 + Y_{t-1} + v_t \quad \text{(DS)},$$

where the $\{v_t\}$'s are usually correlated but stationary. Both (TS) and (DS) give rise to a time series that increases in the mean level over time, but a fundamental difference exists between them. For the first model, a stationary process $\{v_t\}$ results after detrending, while for the second model, a stationary process $\{v_t\}$ results after differencing. If we were to difference the first model, the resulting process becomes

$$\Delta Y_t = \beta_1 + \Delta v_t,$$

so that a noninvertible process $\{\Delta v_t\}$ results. This is undesirable. The real question becomes how we can differentiate between these two models. This is one of the challenging problems for econometricians, and one way to resolve this is to build a model that encompasses both situations. Specifically, consider the model

$$Y_t = \beta_0 + \beta_1 t + v_t,$$

where

$$v_t = \alpha v_{t-1} + Z_t, \quad Z_t \sim N(0, \sigma^2), \quad \text{i.i.d.}$$

Simple algebra shows that

$$
\begin{aligned}
Y_t &= \beta_0 + \beta_1 t + \alpha v_{t-1} + Z_t \\
&= \beta_0 + \beta_1 t + \alpha(Y_{t-1} - \beta_0 - \beta_1(t-1)) + Z_t \\
&= \beta_0(1 - \alpha) + \beta_1(t - \alpha(t-1)) + \alpha Y_{t-1} + Z_t \\
&= \beta_0(1 - \alpha) + \beta_1 \alpha + t\beta_1(1 - \alpha) + \alpha Y_{t-1} + Z_t \\
&:= \gamma_0 + \gamma_1 t + \alpha Y_{t-1} + Z_t,
\end{aligned}
$$

where the symbol := means "defined as." In this case, $\gamma_0 = \beta_0(1 - \alpha) + \beta_1 \alpha$ and $\gamma_1 = \beta_1(1 - \alpha)$. Note that:

1. If $\alpha = 1, \gamma_0 = \beta_1, \gamma_1 = 0$ and $Y_t = \beta_1 + Y_{t-1} + Z_t$, we end up with model (DS), a differenced stationary series.

2. If $\alpha < 1, Y_t = \gamma_0 + \gamma_1 t + \alpha Y_{t-1} + Z_t$, we end up with model (TS), a trend stationary series.

Also observe that the trend stationary $\{Y_t\}$ satisfies

$$Y_t = \gamma_0 + \gamma_1 t + \alpha Y_{t-1} + Z_t,$$

so that

$$(1 - B)Y_t = \gamma_0 + \gamma_1 t + (\alpha - 1)Y_{t-1} + Z_t.$$

We can test for the coefficient $\alpha = 1$. If $\alpha = 1$, recall that $\gamma_0 = \beta_1$ and $\gamma_1 = 0$, so that we have a differenced stationary series

$$(1 - B)Y_t = \beta_1 + Z_t. \tag{8.1}$$

On the other hand, if $\alpha < 1$, we have a trend stationary series

$$(1 - B)Y_t = \gamma_0 + \gamma_1 t + (\alpha - 1)Y_{t-1} + Z_t. \tag{8.2}$$

To perform this test, we have to inspect the regression coefficient $(\alpha - 1)$ of Y_{t-1} in equation (8.2) to test if it is equal to zero. Under the null hypothesis $H : \alpha = 1$, (8.2) reduces to (8.1).

Looking at it differently, if we rewrite (8.2) as

$$Y_t = \gamma_0 + \gamma_1 t + \alpha Y_{t-1} + Z_t, \tag{8.3}$$

then under $H : \alpha = 1$ (recall that $\gamma_0 = \beta_1$ and $\gamma_1 = 0$), (8.3) becomes

$$Y_t = \beta_1 + Y_{t-1} + Z_t. \tag{8.4}$$

Consequently, the problem of testing for the regression coefficient $H : \alpha - 1 = 0$ in (8.2) can be reformulated as the problem of testing for $H : \alpha = 1$ in (8.3), which in turn is reformulated to testing for $H : \alpha = 1$ in the following problem:

$$Y_t = \beta_1 + \alpha Y_{t-1} + Z_t.$$

This problem turns out to be related to the unit root statistic, discussed in the next section.

8.4   UNIT ROOT TEST

The previous discussion leads us to consider the following testing problem. Consider testing for the AR coefficient $H : \alpha = 1$ in the model

$$Y_t = \beta_1 + \alpha Y_{t-1} + Z_t. \tag{8.5}$$

To illustrate the key idea, let us assume further that $\beta_1 = 0$ in equation (8.5). Then $\{Y_t\}$ follows a random walk model under H, and statistical tests for this kind of model are collectively known in the time series and econometric literature as *random walk* or *unit root tests*. Consider the least squares estimate $\hat{\alpha}$ for α in (8.5). It is of the form

$$\hat{\alpha} = \frac{\sum_{t=1}^{n} Y_t Y_{t-1}}{\sum_{t=1}^{n} Y_{t-1}^2}.$$

In particular,

$$n(\hat{\alpha} - 1) = \frac{(1/n) \sum_{t=1}^{n} Y_{t-1} Z_t}{(1/n)^2 \sum_{t=1}^{n} Y_{t-1}^2}. \tag{8.6}$$

To study the properties of the test statistic in equation (8.6), we need to consider the asymptotic behavior of the numerator and denominator. For the denominator, we rely on the simple form of the functional central limit theorem (invariance principle):

Theorem 8.1 *Let $Z_1, \ldots, Z_n$ be a sequence of i.i.d. random variables with mean zero and variance 1. Let $t \in [0, 1]$ be given and let $Y_n(t) = (1/\sqrt{n}) \sum_{i=1}^{[nt]} Z_i$, where $[nt]$ denotes the integer part of the number nt. Then $Y_n(t) \underset{\mathcal{L}}{\to} W(t)$ as n tends to ∞, where $W(t)$ is a standard Brownian motion defined on $[0, 1]$.*

A proof of this theorem can be found in Billingsley (1999). In Section 8.5 we discuss how to use a discretized version of this theorem to simulate the sample path of Brownian motion. In terms of (8.6), observe that under $H : \alpha = 1$, $Y_t = \sum_{i=1}^{t} Z_i$. Therefore, direct application of Theorem 8.1 immediately yields $(1/\sqrt{n})Y(t) \underset{\mathcal{L}}{\to} W(t)$. Accordingly, the denominator in (8.6) can be shown to converge to

$$\frac{1}{n^2}\sum_{t=1}^{n} Y_{t-1}^2 = \sum_{t=1}^{n} \left(\frac{Y_{t-1}}{\sqrt{n}}\right)^2 \frac{1}{n}$$

$$\underset{\mathcal{L}}{\to} \int_0^1 W^2(t)\, dt.$$

Analyzing the numerator in (8.6) is more tricky. Consider the following derivation. Under $H : \alpha = 1$,

$$Y_t = Y_{t-1} + Z_t,$$
$$Y_t^2 = Y_{t-1}^2 + 2Y_{t-1}Z_t + Z_t^2. \tag{8.7}$$

Summing both sides of (8.7) from 1 to n and simplifying yields

$$\sum_{t=1}^{n} Y_{t-1}Z_t = \frac{1}{2}\left(Y_n^2 - \sum_{t=1}^{n} Z_t^2\right). \tag{8.8}$$

Clearly, $Y_n^2/n \underset{\mathcal{L}}{\to} W^2(1)$ and $(1/n)\sum_{t=1}^{n} Z_t^2 \to 1$ almost surely. Substituting these facts into (8.8), we obtain

$$\frac{1}{n}\sum_{t=1}^{n} Y_{t-1}Z_t \underset{\mathcal{L}}{\to} \frac{1}{2}[W^2(1) - 1]$$

$$= \int_0^1 W(t)\, dW(t),$$

where the last step follows from Itô's rule; see, for example, Øksendal (1998). In summary, we have derived the following theorem.

Theorem 8.2 *Let Y_t follow (8.5) with $\beta_1 = 0$. Then under $H : \alpha = 1$,*

$$n(\hat{\alpha} - 1) \underset{\mathcal{L}}{\to} \frac{\int_0^1 W(t)\, dW(t)}{\int_0^1 W^2(t)\, dt}. \tag{8.9}$$

The result above is usually known as the *unit root test statistic* or *Dickey–Fuller statistic*. Its numerical percentiles have been tabulated by various people and can be found in the books of Fuller (1996) and Tanaka (1996). When the assumption that $\beta_1 = 0$ is removed, the argument above can be used with slight modification, and the resulting statistic will be slightly different from (8.9). Details of these extensions can be found in Tanaka (1996).

8.5   SIMULATIONS

As the asymptotic distributions related to unit root tests are mostly non-standard, one has to resort to simulations in empirical studies. To this end, Theorem 9.1 serves as the building block, and one can simulate functionals of Brownian motions as follows.

Consider a discretized version of Theorem 8.1:

$$W(t_{k+1}) = W(t_k) + \epsilon_{t_k}\sqrt{\Delta t}, \qquad (8.10)$$

where $t_{k+1} - t_k = \Delta t$, and $k = 0, \ldots, N$ with $t_0 = 0$. In this equation, $\epsilon_{t_k} \sim N(0,1)$ are i.i.d. random variables. Further, assume that $W(t_0) = 0$. Except for the factor Δt, this equation is the familiar random walk model. Note that from this model, we get for $j < k$,

$$W(t_k) - W(t_j) = \sum_{i=j}^{k-1} \epsilon_{t_i}\sqrt{\Delta t}.$$

There are several consequences:

1. As the right-hand side is a sum of normal random variables, $W(t_k) - W(t_j)$ is also normally distributed.

2. By taking expectations we have

$$E(W(t_k) - W(t_j)) = 0,$$

$$\mathrm{var}\,(W(t_k) - W(t_j)) = E\left(\sum_{i=j}^{k-1} \epsilon_{t_i}\sqrt{\Delta t}\right)^2 = (k-j)\,\Delta t = t_k - t_j.$$

3. For $t_1 < t_2 \le t_3 < t_4$,

$$W(t_4) - W(t_3) \text{ is uncorrelated with } W(t_2) - W(t_1).$$

Equation (8.10) provides a way to simulate a standard Brownian motion (Wiener process). To see how, consider a partition of $[0,1]$ into n subintervals each with length $1/n$. For each number t in $[0,1]$, let $[nt]$ denote the greatest integer part of it. For example, if $n = 10$ and $t = \frac{1}{3}$, then $[nt] = [\frac{10}{3}] = 3$. Now define a stochastic process in $[0,1]$ as follows. For each t in $[0,1]$, define

$$S_{[nt]} = \frac{1}{\sqrt{n}} \sum_{i=1}^{[nt]} \epsilon_i, \qquad (8.11)$$

where ϵ_i are i.i.d. standard normal random variables. Clearly,

$$S_{[nt]} = S_{[nt]-1} + \epsilon_{[nt]}\frac{1}{\sqrt{n}},$$

which is a special form of (8.10) with $\Delta t = 1/n$. Furthermore, we know that at $t = 1$,

$$S_{[nt]} = S_n = \frac{1}{\sqrt{n}} \sum_{i=1}^{n} \epsilon_i$$

has a standard normal distribution. Also by the central limit theorem, we know that S_n tends to a standard normal random variable in distribution even if the ϵ_i are only i.i.d. but not necessarily normally distributed. The idea is that by taking the limit as n tends to ∞, the process $S_{[nt]}$ would tend to a Wiener process in distribution. Consequently, to simulate a sample path of a Wiener process, all we need to do is to iterate equation (8.11).

In other words, by taking limit as Δt tends to zero, we get a Wiener process (Brownian motion):

$$dW(t) = \epsilon(t)\sqrt{dt},$$

where $\epsilon(t)$ are uncorrelated standard normal random variables. We can interpret this equation as a continuous-time approximation of the random walk model (8.10). Of course, the validity of this approximation relies on Theorem 8.1.

8.6 EXERCISES

1. Using equation (8.11), simulate three sample paths of a standard Brownian motion for $n = 100, 500$, and 1000 and plot out these paths with the SPLUS program.

2. Download from the Internet the daily stock prices of any company you like.

   (a) Plot out the daily prices. Can you detect any trend?

   (b) Construct the returns series from these prices and plot the historgram of the returns series. Does it look like a normal curve?

   (c) Construct the log returns series from these prices and plot the histogram of the log returns series. Does it look like a normal curve?

   (d) Use Exercise 1 in conjunction with Theorem 8.2 to simulate the limiting distribution of the unit root statistics.

   (e) Perform a unit root test on the hypothesis that the log returns series of the company follows a random walk model.

9

Heteroskedasticity

9.1 INTRODUCTION

Similar to linear regression analysis, many time series exhibit a heteroskedastic (nonconstant variance) structure. In a linear regression model, if the response variable Y has a nonconstant variance structure such as

$$Y = X\beta + e, \quad \text{where var } (e) = \sigma^2 \begin{pmatrix} w_1 & 0 & \cdots & 0 \\ 0 & w_2 & & 0 \\ \vdots & & \ddots & \vdots \\ 0 & \cdots & & w_n \end{pmatrix},$$

then instead of using the ordinary least squares procedure, we use a generalized least squares (GLS) method to account for the heterogeneity of e. In time series it is often observed that variations of the time series are quite small for a number of successive periods of time, then much larger for a while, then smaller again for apparently no reason. It would be desirable if these changes in variation (volatility) could be incorporated into the model.

A case in point is an asset price series. If a time series of an asset price on a log scale (e.g., stock or bond prices, exchange rates in log scale, etc.) is differenced, the differenced series looks like white noise. In other words, if $\{Y_t\}$ denotes the series of the log of the price of an underlying asset at period t, we usually end up with an ARIMA(0,1,0) model for $\{Y_t\}$,

$$\Delta Y_t = X_t = \epsilon_t, \quad \text{where } \epsilon_t \sim \text{N}(0,1), \quad \text{i.i.d.} \tag{9.1}$$

This equation simply states that the returns of the underlying asset $X_t = Y_t - Y_{t-1}$ behave like a Gaussian white noise sequence, which is consistent with the celebrated random walk hypothesis. Although this seems reasonable as a first-order approximation, further analysis reveals other kinds of structures that cannot be explained adequately by equation (9.1). Typical signatures of these features, usually known as *stylized-facts* are:

1. $\{X_t\}$ is heavy-tailed, much more so than the Gaussian white noise.

2. Although not much structure is revealed in the correlation function of $\{X_t\}$, the series $\{X_t^2\}$ is highly correlated. Sometimes, these correlations are always nonnegative.

3. The changes in $\{X_t\}$ tend to be clustered. Large changes in $\{X_t\}$ tend to be followed by large changes, and small changes in $\{X_t\}$ tend to be followed by small changes.

The last point deserves some attention. The pursuit for understanding changes in the variance or volatility $\{\sigma_t\}$ is important for financial markets; investors require higher expected returns as a compensation for holding riskier assets. Further, a time series with variance changing over time definitely has implications for the validity and efficiency of statistical inference about the parameters. For example, the celebrated Black–Scholes formula used in option pricing requires knowledge of the volatility process $\{\sigma_t\}$.

It is partly for these reasons that time series models with heteroskedastic errors were developed. In this chapter we discuss the popular ARCH and GARCH models briefly. There are other alternatives that can be used to capture the heteroskedastic effects: for example, the stochastic volatility model. A nice review of some recent developments in this field can be found in Shephard (1996). There is also a GARCH module available in the SPLUS program.

9.2   ARCH

One of the earliest time series models for heteroskedasticity is the ARCH model. In its simplest form, an ARCH model expresses the return series X_t as

$$X_t = \sigma_t \epsilon_t,$$

where it is usually assumed that $\epsilon_t \sim N(0, 1)$, i.i.d. and σ_t satisfies

$$\sigma_t^2 = \alpha_0 + \sum_{i=1}^{p} \alpha_i X_{t-i}^2. \tag{9.2}$$

Let $\mathcal{F}_{t-1} = \sigma(X_{t-1}, X_{t-2}, \dots)$ denote the sigma field generated by pass information until time $t-1$. Then

$$E(X_t^2|\mathcal{F}_{t-1}) = E(\sigma_t^2\epsilon_t^2|\mathcal{F}_{t-1})$$
$$= \sigma_t^2 E(\epsilon_t^2|\mathcal{F}_{t-1})$$
$$= \sigma_t^2.$$

This identity implies that the conditional variance of X_t evolves according to previous values of X_t^2 like an AR(p) model: hence the name *pth-order autoregressive conditional heteroskedastic*, ARCH(p), *model*. Clearly, conditions need to be imposed on the coefficients in order to have a well-defined process for equation (9.2). For example, to ensure that $\sigma_t^2 \geq 0$ and X_t is well defined, one sufficient condition is

$$\alpha_i \geq 0 \quad \text{for} \quad i = 0, \dots, p,$$

and

$$\alpha_1 + \cdots + \alpha_p < 1.$$

We do not pursue proof of this statement here. Interested readers can find related discussions in standard references on ARCH [see, e.g., Campbell, Lo, and MacKinlay (1997) or Gouriéroux (1997)]. Instead, let us consider the following ARCH(1) example.

Example 9.1 *Let $X_t = \sigma_t\epsilon_t$ with $\sigma_t^2 = \alpha_0 + \alpha_1 X_{t-1}^2$. Substituting this recursion repeatedly, we have*

$$X_t^2 = \sigma_t^2\epsilon_t^2$$
$$= \epsilon_t^2(\alpha_0 + \alpha_1 X_{t-1}^2)$$
$$= \alpha_0\epsilon_t^2 + \alpha_1 X_{t-1}^2\epsilon_t^2$$
$$= \alpha_0\epsilon_t^2 + \alpha_1\epsilon_t^2(\sigma_{t-1}^2\epsilon_{t-1}^2)$$
$$= \alpha_0\epsilon_t^2 + \alpha_1\epsilon_t^2\epsilon_{t-1}^2(\alpha_0 + \alpha_1 X_{t-2}^2)$$
$$\vdots$$
$$= \alpha_0 \sum_{j=0}^{n} \alpha_1^j\epsilon_t^2\cdots\epsilon_{t-j}^2 + \alpha_1^{n+1}\epsilon_t^2\epsilon_{t-1}^2\cdots\epsilon_{t-n}^2 X_{t-n-1}^2. \qquad \square$$

If $\alpha_1 < 1$, the last term of the expression above tends to zero as n tends to infinity, and in this case,

$$X_t^2 = \alpha_0 \sum_{j=0}^{\infty} \alpha_1^j\epsilon_t^2\cdots\epsilon_{t-j}^2. \qquad (9.3)$$

In particular, $EX_t^2 = \alpha_0/(1-\alpha_1)$. The following conclusions can be deduced from this example.

1. It follows from equation (9.3) that X_t is causal although nonlinear [i.e., it is a nonlinear function of $(\epsilon_t, \epsilon_{t-1}, \dots)$].

2.
$$E(X_t) = E(E(X_t|\mathcal{F}_{t-1})) = E(E(\sigma_t \epsilon_t|\mathcal{F}_{t-1})) = 0.$$

3. var $(X_t) = E(X_t^2) = \alpha_0/(1 - \alpha_1)$.

4. For $h > 0$,
$$
\begin{aligned}
E(X_{t+h}X_t) &= E(E(X_{t+h}X_t|\mathcal{F}_{t+h-1})) \\
&= E(X_t E(\sigma_{t+h}\epsilon_{t+h}|\mathcal{F}_{t+h-1})) \\
&= 0.
\end{aligned}
$$

5. $E(X_t^2|\mathcal{F}_{t-1}) = \sigma_t^2 = \alpha_0 + \alpha_1 X_{t-1}^2$.

Alternatively, we can express the ARCH(1) model as
$$
\begin{aligned}
X_t^2 &= \sigma_t^2 + X_t^2 - \sigma_t^2 \\
&= \alpha_0 + \alpha_1 X_{t-1}^2 + \sigma_t^2(\epsilon_t^2 - 1) \\
&= \alpha_0 + \alpha_1 X_{t-1}^2 + v_t.
\end{aligned}
$$

Formally, we can think of an ARCH(1) as an AR(1) for the process $\{X_t^2\}$ driven by a new noise $\{v_t\}$. By assuming that $0 \le \alpha_1 \le 1$ and the process to be stationary, what can we say about the covariance structure of an ARCH(1)? To answer this question, we have to evaluate the covariance function of X_t^2. To this end, consider
$$
\begin{aligned}
E(\sigma_t^4) &= E(\alpha_0 + \alpha_1 X_{t-1}^2)^2 \\
&= \alpha_0^2 + 2\alpha_1 \alpha_0^2/(1 - \alpha_1) + \alpha_1^2\{\text{var } X_{t-1}^2 + [E(X_{t-1}^2)]^2\}. \quad (9.4)
\end{aligned}
$$
Recall that $X_t^2 = \alpha_0 + \alpha_1 X_{t-1}^2 + v_t$, so that
$$\text{var } (X_t^2) = \text{var } (v_t)/(1 - \alpha_1^2) \quad (9.5)$$
$$
\begin{aligned}
\text{var } (v_t) &= \text{var } (\sigma_t^2(\epsilon_t^2 - 1)) \\
&= E(\sigma_t^4)E(\epsilon_t^2 - 1)^2 \\
&= 2E(\sigma_t^4). \quad (9.6)
\end{aligned}
$$

Substituting equations (9.5) and (9.6) into (9.4), we obtain
$$E(\sigma_t^4) = \alpha_0^2 + 2\alpha_0^2\alpha_1/(1 - \alpha_1) + 2E(\sigma_t^4)\alpha_1^2/(1 - \alpha_1^2) + \alpha_1^2\alpha_0^2/(1 - \alpha_1)^2.$$

Simplifying this expression, we have
$$E(\sigma_t^4)\frac{1 - 3\alpha_1^2}{1 + \alpha_1} = \frac{\alpha_0^2}{1 - \alpha_1}. \quad (9.7)$$

If $1 \geq \alpha_1^2 \geq \frac{1}{3}$, the left-hand side of (9.7) is negative while its right-hand side is positive, leading to a contradiction! Therefore, for the process to be well defined [i.e., $E(\sigma_t^4)$ exists], $\alpha_1^2 \leq \frac{1}{3}$. In this case we can deduce the following observations:

1. For simplicity, suppose that $EX_t^2 = \alpha_0/(1 - \alpha_1) = 1$. Then

$$EX_t^4 = E(\sigma_t^4 \epsilon_t^4) = 3E(\sigma_t^4)$$

$$= 3 \left[\frac{\alpha_0^2}{(1-\alpha_1)^2} \right] \frac{1 - \alpha_1^2}{1 - 3\alpha_1^2}$$

$$= 3 \left(\frac{1 - \alpha_1^2}{1 - 3\alpha_1^2} \right)$$

$$\geq 3 \quad \text{since} \quad \alpha_1^2 \leq \frac{1}{3}.$$

Therefore, the heavy-tailed phenomenon is reflected in this model.

2. Due to the AR(1) structure of $\{X_t^2\}$, it is clear that

$$\text{corr}\,(X_t^2, X_{t-s}^2) = \alpha_1^s \geq 0.$$

This reveals the nonnegative nature of the ACF of X_t^2.

3. The ARCH(1) equation $\sigma_t^2 = \alpha_0 + \alpha_1 X_{t-1}^2$ partially captures the phenomenon that large changes in return would be followed by large changes in variance (volatility).

9.3   GARCH

Similar to the extension from an AR model to an ARMA model, one can extend the notion of an ARCH model to a *generalized* ARCH, (GARCH) *model*. Specifically, a GARCH model can be expressed in the form

$$X_t = \sigma_t \epsilon_t, \quad \epsilon_t \sim \text{N}(0, 1),$$

$$\sigma_t^2 = \alpha_0 + \sum_{i=1}^p \beta_i \sigma_{t-i}^2 + \sum_{j=1}^q \alpha_j X_{t-j}^2. \tag{9.8}$$

Conditions on α's and β's need to be imposed for equation (9.8) to be well defined. Since finding exact conditions for a general GARCH(p, q) model can be tricky, one has to resort to case-by-case study. A glimpse of the technical aspects of this problem can be found in the article by Nelson (1990). One of the main reasons to consider the GARCH extension is that by allowing past volatilities to affect the present volatility in (9.8), a more parsimonious model may result. We shall not pursue the GARCH model in its full generality, but instead, let us consider the following illustrative GARCH(1,1) model.

Example 9.2 *Let X_t be a* GARCH(1,1) *model so that*

$$\sigma_t^2 = \alpha_0 + \alpha_1 X_{t-1}^2 + \beta_1 \sigma_{t-1}^2.$$

Since σ_{t-1}^2 is unobservable, one way is to estimate it from the initial stretch of the data, say the first 50 points. This procedure is often unsatisfactory, and a better approach is to reparametrize it and to think of it as an ARMA *process. Specifically, consider*

$$
\begin{aligned}
X_t^2 &= \sigma_t^2 + (X_t^2 - \sigma_t^2) \\
&= \alpha_0 + \alpha_1 X_{t-1}^2 + \beta_1 \sigma_{t-1}^2 + X_t^2 - \sigma_t^2 \\
&= \alpha_0 + (\alpha_1 + \beta_1) X_{t-1}^2 - \beta_1 (X_{t-1}^2 - \sigma_{t-1}^2) + X_t^2 - \sigma_t^2 \\
&= \alpha_0 + (\alpha_1 + \beta_1) X_{t-1}^2 + v_t - \beta_1 v_{t-1},
\end{aligned}
$$

where $v_t = X_t^2 - \sigma_t^2 = \sigma_t^2(\epsilon_t^2 - 1)$. With this expression, the process $\{X_t^2\}$ can be viewed as an ARMA(1,1) *process driven by the noise v_t.* □

This fact is true in general. One can deduce easily that

Theorem 9.1 *If X_t is a* GARCH(p, q) *process, then X_t^2 is an* ARMA(m, p) *process in terms of $v_t = \sigma_t^2(\epsilon_t^2 - 1)$, where $m = \max\{p, q\}$ with $\alpha_i = 0$, $i > q$ and $\beta_j = 0$, $j > p$.*

Remarks

1. When a GARCH(1,1) model is entertained in practice, it is often found that $\alpha_1 + \beta_1 \cong 1$. When $\alpha_1 + \beta_1 = 1$, the underlying process X_t is no longer stationary and it leads to the name *integrated* GARCH(1,1) [IGARCH(1,1)] *model*. One of the interpretations of the IGARCH(1,1) model is that the volatility is persistent. To see this point clearly, consider

$$
\begin{aligned}
E(\sigma_{t+1}^2|\mathcal{F}_{t-1}) &= E(\alpha_0 + \alpha_1 X_t^2 + \beta_1 \sigma_t^2|\mathcal{F}_{t-1}) \\
&= \alpha_0 + (\alpha_1 + \beta_1)\sigma_t^2 = \alpha_0 + \sigma_t^2. \\
E(\sigma_{t+2}^2|\mathcal{F}_{t-1}) &= E(E(\sigma_{t+2}^2|\mathcal{F}_t)|\mathcal{F}_{t-1}) \\
&= E(\alpha_0 + \sigma_{t+1}^2|\mathcal{F}_{t-1}) \\
&= 2\alpha_0 + \sigma_t^2.
\end{aligned}
$$

In general, by repeating the preceding argument, we have

$$E(\sigma_{t+j}^2|\mathcal{F}_{t-1}) = j\alpha_0 + \sigma_t^2. \tag{9.9}$$

Notice that we have conditioned on $\mathcal{F}_{t-1}$ instead of $\mathcal{F}_t$ since σ_{t+1} is measurable with respect to the smaller sigma field $\mathcal{F}_t$ so that

$$E(\sigma_{t+1}^2|\mathcal{F}_t) = \sigma_{t+1}^2.$$

Therefore, we would like to look at the genuine one-step-ahead prediction $E(\sigma_{t+1}^2|\mathcal{F}_{t-1})$. According to equation (9.9), today's volatility affects the forecast of tomorrow's volatility, and this effect keeps perpetuating into the infinite future. Hence, any shock to X_t^2 or σ_t^2 will be carried forward (i.e., persist). Although the IGARCH(1,1) model bears a strong resemblance to the random walk model, precautions need to be taken for such an analogy; see Nelson (1990) for details.

2. In equation (9.8), only the squares or magnitudes of X_{t-i} and σ_{t-j} affect the current volatility σ_t^2. This is often unrealistic since the market reacts differently to bad news than to good news. There is a certain amount of asymmetry that cannot be explained by (9.8).

3. Owing to the considerations above, there are many generalizations of the GARCH model to capture some of these phenomena. There are t-GARCH, e-GARCH, and n-GARCH models, just to name a few. We shall not discuss any of these here, but the basic concept is to extend the notion of conditional heterogeneity to capture other observed phenomena.

4. The Gaussian assumption of ϵ_t is not crucial. One may relax it to allow for more heavy-tailed distributions, such as a t-distribution. SPLUS does allow one to entertain some of these generalizations. Of course, having the Gaussian assumption facilitates the estimation process.

9.4   ESTIMATION AND TESTING FOR ARCH

According to the definition of ARCH, it is clear that $X_t|\mathcal{F}_{t-1} \sim N(0, \sigma_t^2)$ with conditional probability density function

$$f(x_t|\mathcal{F}_{t-1}) = (2\pi\sigma_t^2)^{-1/2} \exp\left(-\frac{1}{2\sigma_t^2}x_t^2\right),$$

where $\sigma_t^2 = \alpha_0 + \sum_{i=1}^{p} \alpha_i X_{t-i}^2$. By iterating this conditional argument, we obtain

$$f(x_n, \dots, x_1|x_0) = f(x_n|x_{n-1}, \dots, x_0) \cdots f(x_2|x_1, x_0)f(x_1|x_0),$$

$$\log f(x_n, \dots, x_1|x_0) = \sum_{t=1}^{n} \log f(x_t|\mathcal{F}_{t-1})$$

$$= -\frac{n}{2}\log 2\pi + \sum_{t=1}^{n} -\frac{1}{2}\log \sigma_t^2 - \frac{1}{2}\sum_{t=1}^{n}\frac{x_t^2}{\sigma_t^2}. \qquad (9.10)$$

Therefore, substituting $\sigma_t^2(\alpha_0, \alpha_1, \dots, \alpha_p) = \alpha_0 + \sum_{i=1}^{p}\alpha_i X_{t-i}^2$ into the equation above for different values of $(\alpha_0, \dots, \alpha_p)$, the MLEs can be obtained by maximizing the log-likelihood function above numerically. If it turns out that the ϵ_t is not normal, one can still use this method to obtain the pseudo (quasi) maximum likelihood estimate (PMLE or QMLE). Details can be found in Gouriéroux (1997).

Note that this method can also be applied to the GARCH model. Recall that in the linear time series context, it is usually more tricky to estimate an ARMA model than a pure AR model, due to the presence of the MA part. In the estimation of a GARCH model, similar difficulties are encountered. To illustrate the idea, consider the simple GARCH(1,1) example where

$$\sigma_t^2 = \alpha_0 + \alpha_1 X_{t-1}^2 + \beta_1 \sigma_{t-1}^2. \tag{9.11}$$

Rewriting the equation above in terms of the volatility σ_t^2 yields

$$\begin{aligned}
\sigma_t^2 &= (1 - \beta_1 B)^{-1}(\alpha_0 + \alpha_1 X_{t-1}^2) \\
&= (1 + \beta_1 B + \beta_1^2 B^2 + \cdots)(\alpha_0 + \alpha_1 X_{t-1}^2).
\end{aligned}$$

Therefore, the volatility σ_t^2 depends on all the past values of X_t^2. To evaluate σ_t^2 from this equation, we truncate the values of $X_t = 0$ for $t \leq 0$ and $\sigma_t^2 = 0$ for $t \leq 0$. Then σ_t^2 is approximated by $\tilde{\sigma}_t^2$ for $t = 1, 2, \ldots$ recursively as

$$\tilde{\sigma}_t^2 = \alpha_0 + \alpha_1 \tilde{X}_{t-1}^2 + \beta_1 \tilde{\sigma}_{t-1}^2,$$

where $\tilde{X}_t = 0$ for $t \leq 0$, $\tilde{X}_t = X_t$ for $t > 0$, and $\tilde{\sigma}_t^2 = 0$ for $t \leq 0$. By iterating this equation, we get

$$\begin{aligned}
\tilde{\sigma}_1^2 &= \alpha_0, \\
\tilde{\sigma}_2^2 &= \alpha_0 + \alpha_1 \tilde{X}_1^2 + \beta_1 \tilde{\sigma}_1^2 \\
&= \alpha_0 + \alpha_1 X_1^2 + \beta_1 \alpha_0,
\end{aligned}$$

$$\vdots$$

Substituting this expression for $\tilde{\sigma}_t^2$ for various values of $\theta = (\alpha_0, \alpha_1, \beta_1)$ into the likelihood function (9.10) and maximizing it, we can find the MLE numerically. In general, the same idea applies to a GARCH(p, q) model.

Finally, another popular approach to estimation is the generalized method of moments or the efficient method of moments, details of which can be found in Shephard (1996).

As far as testing is concerned, there are many methods. In what follows, we discuss three simple approaches.

1. *Time series test.* Since an ARCH(p) implies that $\{X_t^2\}$ follows an AR(p), one can use the Box–Jenkins approach to study the correlation structure of X_t^2 to identify the AR properties.

2. *Portmanteau tests for residuals.* Let the Portmanteau statistic be defined as

$$Q = n(n+2) \sum_{j=1}^{h} r^2(j)/(n-j),$$

where $r(j)$ denotes the ACF of the fitted residual of X_t^2. If the model is specified correctly, then for large sample sizes,

$$Q \underset{\mathcal{L}}{\to} \chi^2_{h-m},$$

where m is the number of independent parameters used in the model and $\underset{\mathcal{L}}{\to}$ denotes convergence in distribution as the sample size $n \to \infty$.

3. *Lagrange multiplier test.*

Theorem 9.2 *Assuming that X_t is an* ARCH(p) *model, regress X_t^2 with respect to $X_{t-1}^2, \ldots, X_{t-p}^2$ and form the fitted regression equation*

$$\hat{X}_t^2 = \hat{\alpha}_0 + \sum_{i=1}^{p} \hat{\alpha}_i X_{t-i}^2. \tag{9.12}$$

Let R^2 denote the coefficient of determination from (9.12). Then under the null hypothesis $H: \alpha_1 = \cdots = \alpha_p = 0$ (i.e., no heterogeneity),

$$nR^2 \underset{\mathcal{L}}{\to} \chi_p^2.$$

Note that Theorem 9.2 is stated in terms of nR^2, which can be shown to be asymptotically equivalent to the Lagrange multiplier test statistic. Again, details of this equivalence can be found in Gouriéroux (1997). Although this is a widely used statistics, it has been documented that it has relatively low power. Thus we should use it with caution. When the underlying conditional structure is misspecified, rejecting the null hypothesis may not necessarily mean that the GARCH effect exists.

9.5   EXAMPLE OF FOREIGN EXCHANGE RATES

In this example, the weekly exchange rates of the U.S. dollar and British pound sterling between the years 1980 and 1988 analyzed earlier are studied again. The data are stored in the file `exchange.dat` on the Web site for this book. Let `ex.s` denote the SPLUS object that contains the data set. We perform a time series analysis on `ex.s` as follows. First, the histogram of the data is plotted and it is compared with the normal density. As usual, we start by differencing the data.

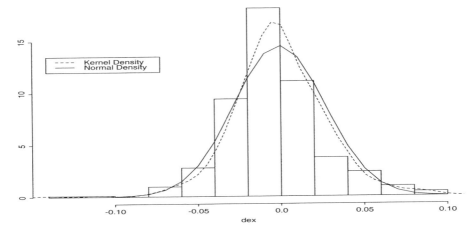

Fig. 9.1 Histograms of the differenced exchange rates.

```
>dex_diff(ex.s)
>hist(dex, prob=T, col=0)
>library(MASS)
>c(ucv(dex,0),bcv(dex,0))
[1]   0.0038 0.032
(This command gives an estimate for the size of the window
 used in constructing a kernel density.)
>lines(density(dex,width=0.03),lty=3)
(This command draws a kernel estimate of the histogram of
the data with a bin width of 0.03)
>x_seq(-0.1,0.1,0.01)
>lines(x,dnorm(x,mean(dex),sqrt(var(dex))),lty=1)
(This command draws a normal density with mean and variance
being the mean and variance of the data, i.e., treating the data
as normally distributed.)
> leg.names_c("Kernel Density", "Normal Density")
> legend(locator(1), leg.names, lty=c(3,1))
(These last two commands produce the legends for the graphs.)
```

If we look at Figure 9.1, the heavy-tailed phenomenon, particularly the right-hand tail, is clear. Next, an exploratory time series analysis is performed to compare the structures of the ACF of the data, the differenced data, and the squares of the differenced data.

```
>par(mfrow=c(3,2))
>tsplot(ex.s)
>acf(ex.s)
>tsplot(dex)
```

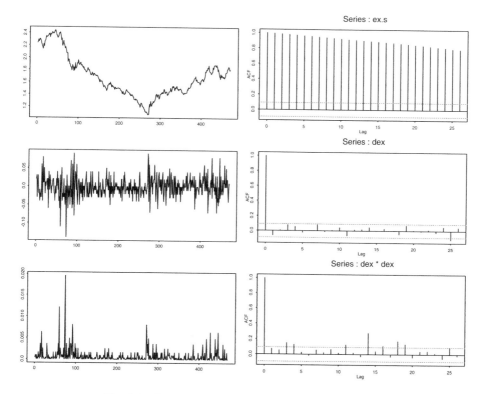

Fig. 9.2 Time series plots and ACFs for X_t and X_t^2.

```
>acf(dex)
>tsplot(dex*dex)
>acf(dex*dex)
```

Inspection of Figure 9.2 confirms the possibility of an ARCH/GARCH structure. To test for this possibility, we use the Lagrange multiplier (LM) test. To this end, regress X_t^2 with respect to $X_{t-1}^2, X_{t-2}^2, X_{t-3}^2, X_{t-4}^2$. Lag 4 is chosen because analysis with further lags does not reveal extra information.

```
>x1_lag(dex)
>x1_c(0,x1[1:468])
>x2_lag(x1)
>x2_c(0,x2[1:468])
>x3_lag(x2)
>x3_c(0,x3[1:468])
>x4_lag(x3)
>x4_c(0,x4[1:468])
>z1_x1*x1
>z2_x2*x2
```

```
>z3_x3*x3
>z4_x4*x4
>lm.1_lm(dex*dex~z1+z2+z3+z4)
>summary(lm.1)
```

```
Call: lm(formula = dex * dex ~ z1 + z2 + z3 + z4)
Residuals:
      Min            1Q       Median           3Q       Max
 -0.002848  -0.0006073  -0.0004331  0.0001337   0.0187
```

```
Coefficients:
              Value Std. Error  t value  Pr(>|t|)
(Intercept) 0.0005 0.0001        5.4863   0.0000
         z1 0.0469 0.0461        1.0171   0.3096
         z2 0.0333 0.0457        0.7275   0.4673
         z3 0.1379 0.0458        3.0130   0.0027
         z4 0.1099 0.0461        2.3813   0.0177
```

```
Residual standard error: 0.00151 on 464 degrees of freedom
Multiple R-Squared: 0.04026
F-statistic: 4.866 on 4 and 464 degrees of freedom, the
             p-value is 0.0007485
```

```
Correlation of Coefficients:
    (Intercept)       z1       z2       z3
z1 -0.2837
z2 -0.3088       -0.0634
z3 -0.3065       -0.0387 -0.0682
z4 -0.2807       -0.1450 -0.0389 -0.0655
```

```
> t_470*0.04026
> 1-pchisq(t,4)
[1] 0.0008140926
```

The p-value of the LM test is 8×10^{-4}, and consequently, the presence of heterogeneity is confirmed. The last step is to fit a GARCH model to the data.

```
>module(garch)
>dex.mod_garch(dex~-1,~garch(1,1))
> summary(dex.mod)
```

```
Call: garch(formula.mean = dex ~ -1, formula.var = ~ garch(1,1))
```

```
Mean Equation: dex ~ -1
```

```
Conditional Variance Equation:   ~ garch(1, 1)

Conditional Distribution:  gaussian

----------------------------------------------------------------

Estimated Coefficients:
----------------------------------------------------------------
              Value Std.Error t value   Pr(>|t|)
        A 3.157e-05 1.374e-05   2.298 0.0110107
  ARCH(1) 6.659e-02 2.045e-02   3.256 0.0006068
 GARCH(1) 8.915e-01 3.363e-02  26.511 0.0000000


----------------------------------------------------------------

AIC(3) = -2064.388
BIC(3) = -2051.936
...
>plot(dex.mod)
```

From this fit, we arrive at the model

$$\sigma_t^2 = 3.2 \times 10^{-5} + 0.89\sigma_{t-1}^2 + 0.067X_{t-1}^2.$$

Figure 9.3 displays the plot of the fitted model, which consists of the ACF of the squares of the data, the conditional standard deviations, the ACF of the residuals of the fitted model, the normal probability (Q-Q norm) plot of the residuals, and so on. There are other options from the SPLUS program; readers can experiment with them. From the Q-Q norm plot, heavy-tailed effects still exist in the residual. Further, $\hat{\alpha}_1 + \hat{\beta}_1 \cong 1$; a possible IGARCH(1,1) model is considered. We can do this by executing the following commands:

```
>new.mod_revise(dex.mod)
Make a selection (or 0 to exit):

1: edit: C
2: edit: CO
3: edit: arch
4: edit: garch
Selection: 3
```

A data window such as Figure 9.4 comes up at this point. For example, if you want to change the α_1 value in UNIX, highlight it at the top row and hit the delete key. Type in the new value you want, 0.069, and hit the return key for this example. This will change the arch.value from the original display into 0.069. If you would like to use this as a starting value and estimate it from the model again, leave the default value under arch.which as 1. If you

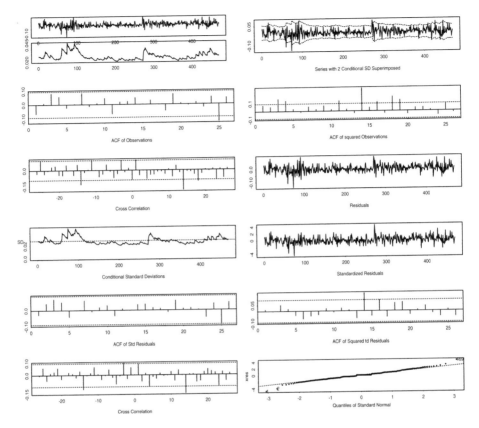

Fig. 9.3   Plots of `dex.mod`.

would like to fix it, change this value to 0. In this example, we leave it as 1. Click on the commit and quit icons to commit the change.

Suppose now that we would like to set $\alpha_0 = 0$. Choose 2 from the edit menu. Another data menu like Figure 9.5 comes up. Change the value to 0 and the `which` level to 0 to fix it. Click on the commit and quit icons again. The model has now been revised. Now fit the revised model with `garch`.

Fig. 9.4   Data editor.

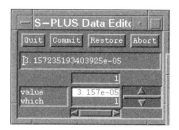

Fig. 9.5 Data editor.

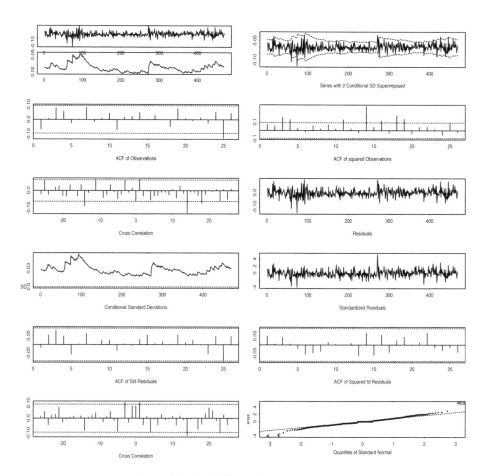

Fig. 9.6 Plots of `dex1.mod`.

```
>dex1.mod_garch(series=dex,model=new.mod)
>plot(dex1.mod)
```

Essentially, we arrive at the revised model:

$$\sigma_t^2 = 0.069X_{t-1}^2 + 0.93\sigma_{t-1}^2.$$

This is an IGARCH(1,1) with plots given in Figure 9.6. This seems to provide a better fit than the GARCH(1,1) alone in the sense that the Q-Q norm plot has a less heavy-tailed effect. The improvement seems to be marginal, however.

There are other kinds of options that one can try with the SPLUS program. For example, one can simulate the fitted model, change the distribution of the error term to nonnormal, predict future σ_t, and so on. We shall not pursue these details; interested readers can explore these options with the `help` command or consult the SPLUS GARCH manual.

9.6 EXERCISES

1. Let X_t be an ARCH(1) process

$$\sigma_t^2 = \alpha_0 + \alpha_1 X_{t-1}^2.$$

(a) Show that

$$E(\sigma_t^4) = \frac{\alpha_0^2}{1 - \alpha_1} \frac{1 + \alpha_1}{1 - 3\alpha_1^2}.$$

(b) Deduce that

$$E(X_t^4) = 3 \left(\frac{1 - \alpha_1^2}{1 - 3\alpha_1^2} \right).$$

2. Show that for an IGARCH(1,1) model, for $j \geq 0$,

$$E(\sigma_{t+s}^2 | \mathcal{F}_{t-1}) = j\alpha_0 + \sigma_t^2.$$

3. Let X_t be a GARCH(2,3) model,

$$\sigma_t^2 = \alpha_0 + \sum_{i=1}^{2} \beta_i \sigma_{t-i}^2 + \sum_{j=1}^{3} \alpha_j X_{t-j}^2.$$

Show that X_t^2 can be written as an ARMA(3,2) model in terms of the process $\nu_t = \sigma_t^2(\epsilon_t^2 - 1)$. Identify the parameters of the ARMA process in terms of the parameters of the given GARCH(2,3) model.

4. Perform a GARCH analysis on the U.S. Treasury bill data set discussed in Chapter 6.

10

Multivariate Time Series

10.1 INTRODUCTION

Often, time series arising in practice are best considered as components of
some vector-valued (multivariate) time series $\{\boldsymbol{X}_t\}$ having not only serial
dependence within each component series $\{X_{ti}\}$, but also interdependence
between different components $\{X_{ti}\}$ and $\{X_{tj}\}$, $i \neq j$. Much of the theory
of univariate time series extends in a natural way to the multivariate case;
however, two important problems arise.

1. *Curse of dimensionality.* As the number of components in $\{\boldsymbol{X}_t\}$ increases, the number of parameters increases. As an example, for a multivariate time series consisting of a portfolio of 10 equities, even a simple
   vector AR(1) model may need up to 100 freely varying parameters; the
   curse of dimensionality comes into effect.

2. *Identifiability.* Contrary to the univariate case, it is not true that an
   arbitrary vector ARMA model can be identified uniquely. We shall see
   an example of this later.

In this chapter we aim to provide a brief summary of multiple time series
which leads to the development of vector autoregressive and moving average
(VARMA) models. We do not attempt to provide a comprehensive account of
the details of multiple time series; interested readers are referred to Lütkepohl
(1993) and Brockwell and Davis (1991).

First introduce a few basic notions. A *k-variate time series* is a stochastic
process consisting of k-dimensional random vectors $(X_{t1}, X_{t2}, \ldots, X_{tk})'$ ob-

served at times t (usually, $t = 1, 2, \dots$). The component series $\{X_{ti}\}$ could be studied independently as univariate time series, each characterized, from a second-order point of view, by its own mean and autocovariance function. Such an approach, however, fails to take into account the possible dependence *between* the two component series, and such cross-dependence may be of great importance for predicting future values of the component series.

Consider the series of random vectors, $\boldsymbol{X}_t = (X_{t1}, \dots, X_{tk})'$ and define the mean vector

$$\boldsymbol{\mu}_t = E(\boldsymbol{X}_t) = (E(X_{t1}), \dots, E(X_{tk}))'$$

and covariance matrices

$$\boldsymbol{\Gamma}(t+h, t) = \mathrm{cov}(\boldsymbol{X}_{t+h}, \boldsymbol{X}_t) = \begin{pmatrix} \gamma_{11}(t+h, t) & \cdots & \gamma_{1k}(t+h, t) \\ \vdots & & \vdots \\ \gamma_{k1}(t+h, t) & \cdots & \gamma_{kk}(t+h, t) \end{pmatrix},$$

where $\gamma_{ij}(t+h, t) = \mathrm{cov}(X_{t+h,i}, X_{t,j})$. In matrix notation,

$$\boldsymbol{\Gamma}(t+h, t) = E(\boldsymbol{X}_{t+h} - \boldsymbol{\mu}_{t+h})(\boldsymbol{X}_t - \boldsymbol{\mu}_t)'.$$

The series $\{\boldsymbol{X}_t\}$ is said to be *stationary* if the moments $\boldsymbol{\mu}_t$ and $\boldsymbol{\Gamma}(t+h, t)$ are both independent of t, in which case we use the notation

$$\boldsymbol{\mu} = E(\boldsymbol{X}_t)$$

and

$$\boldsymbol{\Gamma}(h) = \mathrm{cov}(\boldsymbol{X}_{t+h}, \boldsymbol{X}_t).$$

The diagonal elements of the matrix above are the autocovariance functions of the univariate series $\{X_{ti}\}$, while the off-diagonal elements are the covariances between $X_{t+h,i}$ and X_{tj}, $i \neq j$. Notice that $\gamma_{ij}(h) = \gamma_{ji}(-h)$. Correspondingly, the autocorrelation matrix is defined as

$$R(h) = \begin{pmatrix} \rho_{11}(h) & \cdots & \rho_{1k}(h) \\ \vdots & & \vdots \\ \rho_{k1}(h) & \cdots & \rho_{kk}(h) \end{pmatrix},$$

where

$$\rho_{ij}(h) = \gamma_{ij}(h)(\gamma_{ii}(0)\gamma_{jj}(0))^{-1/2}.$$

Example 10.1 *Consider the bivariate time series*

$$X_{t1} = Z_t,$$

$$X_{t2} = Z_t + 0.75Z_{t-10}, \quad Z_t \sim \mathrm{WN}(0, 1).$$

Then $\boldsymbol{\mu} = 0$ and

$$\boldsymbol{\Gamma}(0) = E\left[\begin{pmatrix} Z_t \\ Z_t + 0.75Z_{t-10} \end{pmatrix}(Z_t, Z_t + 0.75Z_{t-10})\right] = \begin{pmatrix} 1 & 1 \\ 1 & 1 + (0.75)^2 \end{pmatrix}.$$

Similarly,

$$\mathbf{\Gamma}(-10) = \begin{pmatrix} 0 & 0.75 \\ 0 & 0.75 \end{pmatrix} \quad and \quad \mathbf{\Gamma}(10) = \begin{pmatrix} 0 & 0 \\ 0.75 & 0.75 \end{pmatrix}.$$

Further,

$$R(-10) = \begin{pmatrix} 0 & 0.6 \\ 0 & 0.48 \end{pmatrix}, \quad R(0) = \begin{pmatrix} 1 & 0.8 \\ 0.8 & 1 \end{pmatrix},$$

and

$$R(10) = \begin{pmatrix} 0 & 0 \\ 0.6 & 0.48 \end{pmatrix} \qquad \qquad \square$$

Theorem 10.1 *The following results hold:*

1. $\mathbf{\Gamma}(h) = \mathbf{\Gamma}(-h)'.$

2. $|\gamma_{ij}(h)| \leq [\gamma_{ii}(0)\gamma_{jj}(0)]^{1/2}.$

3.

$$\sum_{i,j=1}^{n} \mathbf{a}_i' \mathbf{\Gamma}(i-j) \mathbf{a}_j \geq 0,$$

   for $n = 1, 2, \ldots$ *and* $\mathbf{a}_1, \mathbf{a}_2, \ldots, \mathbf{a}_n \in R^k.$

Proof

1. By assuming that $\boldsymbol{\mu} = \mathbf{0}$,

$$\begin{aligned}
\mathbf{\Gamma}(h) &= E(\mathbf{X}_{t+h}\mathbf{X}_t') \\
&= E(\mathbf{X}_t \mathbf{X}_{t+h}')' \\
&= E(\mathbf{X}_{t-h}\mathbf{X}_t')' \quad \text{by stationarity} \\
&= \mathbf{\Gamma}(-h)'.
\end{aligned}$$

2. Follows from $|\rho_{ij}(h)| \leq 1$.

3. Follows from

$$E\left(\sum_{j=1}^{n} \mathbf{a}'(\mathbf{X}_j - \boldsymbol{\mu})\right)^2 \geq 0.$$

This is also known as the *positive semidefinite property* of $\mathbf{\Gamma}(\cdot)$. $\qquad \square$

The simplest multivariate time series is a white noise.

Definition 10.1 $\{\boldsymbol{Z}_t\} \sim \mathrm{WN}(\boldsymbol{0}, \boldsymbol{\Sigma})$ *if and only if* $\{\boldsymbol{Z}_t\}$ *is stationary with* $\boldsymbol{\mu} = \boldsymbol{0}$ *and*

$$\boldsymbol{\Gamma}(h) = \begin{cases} \boldsymbol{\Sigma} & \text{if } h = 0, \\ \boldsymbol{0} & \text{otherwise.} \end{cases}$$

Definition 10.2 $\{\boldsymbol{Z}_t\} \sim \mathrm{i.i.d.}(\boldsymbol{0}, \boldsymbol{\Sigma})$ *if* $\{\boldsymbol{Z}_t\}$ *are independent identically distributed with* $\boldsymbol{\mu} = \boldsymbol{0}$ *and covariance matrix* $\boldsymbol{\Sigma}$.

Definition 10.3 $\{\boldsymbol{X}_t\}$ *is a linear process if it can be expressed as*

$$\boldsymbol{X}_t = \sum_{j=-\infty}^{\infty} \boldsymbol{C}_j \boldsymbol{Z}_{t-j} \quad \text{for } \{\boldsymbol{Z}_t\} \sim \mathrm{WN}(\boldsymbol{0}, \boldsymbol{\Sigma}),$$

where $\{\boldsymbol{C}_j\}$ *is a sequence of* $k \times k$ *matrices whose entries are absolutely summable, that is,*

$$\sum_{j=-\infty}^{\infty} |\boldsymbol{C}_j(i,l)| < \infty \quad \text{for } i, l = 1, 2, \ldots, k.$$

Clearly, for a linear process $\boldsymbol{X}_t$, $E(\boldsymbol{X}_t) = \boldsymbol{0}$ and

$$\boldsymbol{\Gamma}(h) = \sum_{j=-\infty}^{\infty} \boldsymbol{C}_{j+h} \boldsymbol{\Sigma} \boldsymbol{C}_j', \quad h = 0, \pm 1, \pm 2, \ldots.$$

An MA(∞) representation is the case where $\boldsymbol{C}_j = \boldsymbol{0}$ for $j < 0$,

$$\boldsymbol{X}_t = \sum_{j=0}^{\infty} \boldsymbol{C}_j \boldsymbol{Z}_{t-j}.$$

An AR(∞) representation is such that

$$\boldsymbol{X}_t + \sum_{j=1}^{\infty} \boldsymbol{A}_j \boldsymbol{X}_{t-j} = \boldsymbol{Z}_t.$$

Finally, for a spectral representation, if $\sum_h |\gamma_{ij}(h)| < \infty$ for every $i, j = 1, 2$, then

$$f(\lambda) = \frac{1}{2\pi} \sum_{h=-\infty}^{\infty} e^{-i\lambda h} \boldsymbol{\Gamma}(h), \quad -\pi \le \lambda \le \pi,$$

and $\boldsymbol{\Gamma}$ can be expressed in terms of

$$\boldsymbol{\Gamma}(h) = \int_{-\pi}^{\pi} e^{i\lambda h} f(\lambda) \, d\lambda.$$

Note that both $\boldsymbol{\Gamma}(h)$ and $f(\lambda)$ are $k \times k$ matrices in this case.

10.2   ESTIMATION OF μ AND Γ

A natural estimator of the mean vector μ in terms of the observations $X_1, X_2, \ldots, X_n$, is the vector of sample means

$$\bar{X} = \frac{1}{n} \sum_{t=1}^{n} X_t.$$

For $\Gamma(h)$ we use

$$\hat{\Gamma}(h) = \begin{cases} \dfrac{1}{n} \displaystyle\sum_{t=1}^{n-h} (X_{t+h} - \bar{X})(X_t - \bar{X})' & \text{for } 0 \leq h \leq n-1, \\ \hat{\Gamma}(h)' & \text{for } -n+1 \leq h < 0. \end{cases}$$

The correlation $\rho_{ij}(h)$ between $X_{t+h,i}$ and $X_{t,j}$ is estimated by

$$\hat{\rho}_{ij}(h) = \hat{\gamma}_{ij}(h)(\hat{\gamma}_{ii}(0)\hat{\gamma}_{jj}(0))^{-1/2},$$

where $\hat{\gamma}_{ij}(h)$ is the (i,j)th element of $\hat{\Gamma}(h)$. If $i = j$, $\hat{\rho}_{ij}$ reduces to the sample autocorrelation function of the ith series.

10.3   MULTIVARIATE ARMA PROCESSES

As in the univariate case, we can define an extremely useful class of multivariate stationary processes $\{X_t\}$ by requiring that $\{X_t\}$ should satisfy a set of linear difference equations with constant coefficients. The multivariate white noise $\{Z_t\}$ constitutes the fundamental building block for constructing vector ARMA processes.

Definition 10.4 *$\{X_t\}$ is an* ARMA(p,q) *process if $\{X_t\}$ is stationary and if for every t,*

$$X_t - \Phi_1 X_{t-1} - \cdots - \Phi_p X_{t-p} = Z_t + \Theta_1 Z_{t-1} + \cdots + \Theta_q Z_{t-q},$$

where $\{Z_t\} \sim$ WN$(0, \Sigma)$. [$\{X_t\}$ is an ARMA(p,q) *process with mean μ if $\{X_t - \mu\}$ is an* ARMA(p,q) *process].*

Equivalently, we can write this as

$$\Phi(B)X_t = \Theta(B)Z_t, \quad \{Z_t\} \sim \text{WN}(0, \Sigma),$$

where $\Phi(z) = I_k - \Phi_1 z - \cdots - \Phi_p z^p$ and $\Theta(z) = I_k + \Theta_1 z + \cdots + \Theta_q z^q$ are matrix-valued polynomials. Note that each component of the matrices $\Phi(z)$ and $\Theta(z)$ is a polynomial in z with real coefficients and degree less than p, q, respectively.

Example 10.2 *Setting $p = 1$ and $q = 0$ gives the defining equation*

$$X_t = \Phi_1 X_{t-1} + Z_t$$

of a multivariate AR(1) *series* $\{X_t\}$. *By using a similar argument in the univariate case, we can express* X_t *as*

$$X_t = \sum_{j=0}^{\infty} \Phi_1^j Z_{t-j}$$

if all the eigenvalues of Φ_1 *are less than 1 in absolute value, that is,*

$$\det(I - z\Phi_1) \neq 0 \quad \text{for all } z \in C \text{ such that } |z| \leq 1.$$

In this case we have an MA(∞) *representation with* $C_j = \Phi_1^j$.   □

10.3.1   Causality and Invertibility

Definition 10.5 (Causality) *An* ARMA(p, q) *process* $\Phi(B)X_t = \Theta(B)Z_t$ *is said to be* **causal** *if there exist matrices* $\{\Psi_j\}$ *with absolutely summable components such that*

$$X_t = \sum_{j=1}^{\infty} \Psi_j Z_{t-j} \quad \text{for all } t.$$

Causality is equivalent to the condition

$$\det(\Phi(z)) \neq 0 \quad \text{for all } z \in C \text{ such that } |z| \leq 1.$$

The matrices Ψ_j are found recursively from the equations

$$\Psi_j = \Theta_j + \sum_{k=1}^{\infty} \Phi_k \Psi_{j-k}, \quad j = 0, 1, \dots,$$

where we define $\Theta_0 = I_k$, $\Theta_j = 0$ for $j > q$, $\Phi_j = 0$ for $j > p$ and $\Psi_j = 0$ for $j < 0$.

Definition 10.6 (Invertibility) *An* ARMA(p, q) *process* $\Phi(B)X_t = \Theta(B)Z_t$ *is said to be* **invertible** *if there exist matrices* $\{\Pi_j\}$ *with absolutely summable components such that*

$$Z_t = \sum_{j=0}^{\infty} \Pi_j X_{t-j} \quad \text{for all } t.$$

Invertibility is equivalent to the condition

$$\det(\Theta(z)) \neq 0 \quad \text{for all } z \in C \text{ such that } |z| \leq 1.$$

The matrices $\mathbf{\Pi}_j$ are found recursively from the equations

$$\mathbf{\Pi}_j = -\mathbf{\Phi}_j - \sum_{k=1}^{\infty} \mathbf{\Theta}_k \mathbf{\Pi}_{j-k}, \quad j = 0, 1, \dots,$$

where we define $\mathbf{\Phi}_0 = -\mathbf{I}_k$, $\mathbf{\Phi}_j = 0$ for $j > p$, $\mathbf{\Theta}_j = \mathbf{0}$ for $j > q$, and $\mathbf{\Pi}_j = \mathbf{0}$ for $j < 0$.

10.3.2   Identifiability

For the multivariate AR(1) process in Example 10.2 with

$$\mathbf{\Phi}_1 = \begin{pmatrix} 0 & 0.5 \\ 0 & 0 \end{pmatrix},$$

we get

$$\mathbf{X}_t = \sum_{j=0}^{\infty} \mathbf{\Phi}_1^j \mathbf{Z}_{t-j},$$

so $\mathbf{\Psi}_j = \mathbf{\Phi}^j$.

$$\begin{pmatrix} X_{t1} \\ X_{t2} \end{pmatrix} = \begin{pmatrix} 0 & 0.5 \\ 0 & 0 \end{pmatrix} \begin{pmatrix} X_{t-1,1} \\ X_{t-1,2} \end{pmatrix} + \begin{pmatrix} Z_{t1} \\ Z_{t2} \end{pmatrix}$$

$$= \begin{pmatrix} 0.5X_{t-1,2} + Z_{t1} \\ Z_{t2} \end{pmatrix} = \begin{pmatrix} Z_{t1} \\ Z_{t2} \end{pmatrix} + \begin{pmatrix} 0 & 0.5 \\ 0 & 0 \end{pmatrix} \begin{pmatrix} Z_{t-1,1} \\ Z_{t-1,2} \end{pmatrix}$$

$$= \mathbf{Z}_t + \mathbf{\Phi}_1 \mathbf{Z}_{t-1}$$

[i.e., $\{\mathbf{X}_t\}$ has an alternative representation as an MA(1) process]. As a conclusion, the data can be represented as either a VARMA(1,0) or a VARMA(0,1) model.

More generally, consider a VARMA(1,1) model of the form

$$\mathbf{X}_t = \mathbf{\Phi} \mathbf{X}_{t-1} + \mathbf{Z}_t + \mathbf{\Theta} \mathbf{Z}_{t-1},$$

where $\mathbf{\Phi} = \begin{pmatrix} 0 & \alpha + m \\ 0 & 0 \end{pmatrix}$ and $\mathbf{\Theta} = \begin{pmatrix} 0 & -m \\ 0 & 0 \end{pmatrix}$. It can easily be seen that $(I - \mathbf{\Phi}B)^{-1} = (I + \mathbf{\Phi}B)$, so that the MA representation of $\mathbf{X}_t$ is

$$\mathbf{X}_t = (I - \mathbf{\Phi}B)^{-1}(I + \mathbf{\Theta}B)\mathbf{Z}_t$$

$$= (I + \mathbf{\Phi}B)(I + \mathbf{\Theta}B)\mathbf{Z}_t$$

$$= \mathbf{Z}_t + \begin{pmatrix} 0 & \alpha \\ 0 & 0 \end{pmatrix} \mathbf{Z}_{t-1}.$$

For any given value of m, the given VARMA(1,1) model leads to the preceding MA(1) representation. In other words, this MA(1) equation corresponds to an infinite number of VARMA(1,1) models with $\boldsymbol{\Phi} = \begin{pmatrix} 0 & \alpha + m \\ 0 & 0 \end{pmatrix}$ and

$\boldsymbol{\Theta} = \begin{pmatrix} 0 & -m \\ 0 & 0 \end{pmatrix}$. Moreover, each of these VARMA(1,1) models is always causal and invertible. Consequently, it is not always true that we can identify a VARMA model uniquely from a given MA(∞) representation.

Further restrictions need to be imposed, and details on these issues are beyond the scope of this chapter. Interested readers may find further discussion in Lütkepohl (1993). From now on we shall assume that a convenient form of a VARMA model has been established and we shall proceed with our analysis based on this form.

10.4 VECTOR AR MODELS

Consider a vector AR (or VAR) time series model

$$\boldsymbol{X}_t = \boldsymbol{\nu} + \boldsymbol{\Phi}_1 \boldsymbol{X}_{t-1} + \ldots + \boldsymbol{\Phi}_p \boldsymbol{X}_{t-p} + \boldsymbol{Z}_t,$$

with $\boldsymbol{Z}_t \sim \text{WN}(\boldsymbol{0}, \boldsymbol{\Sigma})$. Note that $\boldsymbol{\nu} = (\nu_1, \ldots, \nu_k)'$ a fixed vector of intercept terms allowing for the possibility of a nonzero mean $E(\boldsymbol{X}_t)$. First consider a VAR(1),

$$\boldsymbol{X}_t = \boldsymbol{\nu} + \boldsymbol{\Phi}_1 \boldsymbol{X}_{t-1} + \boldsymbol{Z}_t.$$

If this generation mechanism starts at time $t = 1$, we get

$$\boldsymbol{X}_1 = \boldsymbol{\nu} + \boldsymbol{\Phi}_1 \boldsymbol{X}_0 + \boldsymbol{Z}_1$$
$$\boldsymbol{X}_2 = \boldsymbol{\nu} + \boldsymbol{\Phi}_1 \boldsymbol{X}_1 + \boldsymbol{Z}_2 = \boldsymbol{\nu} + \boldsymbol{\Phi}_1 (\boldsymbol{\nu} + \boldsymbol{\Phi}_1 \boldsymbol{X}_0 + \boldsymbol{Z}_1) + \boldsymbol{Z}_2$$
$$= (\boldsymbol{I}_k + \boldsymbol{\Phi}_1)\boldsymbol{\nu} + \boldsymbol{\Phi}_1^2 \boldsymbol{X}_0 + \boldsymbol{\Phi}_1 \boldsymbol{Z}_1 + \boldsymbol{Z}_2$$
$$\vdots$$
$$\boldsymbol{X}_t = (\boldsymbol{I}_k + \boldsymbol{\Phi}_1 + \cdots + \boldsymbol{\Phi}_1^{t-1})\boldsymbol{\nu} + \boldsymbol{\Phi}_1^t \boldsymbol{X}_0 + \sum_{i=0}^{t-1} \boldsymbol{\Phi}_1^i \boldsymbol{Z}_{t-i}.$$

Continuing with this process into the remote pass, we may write the VAR(1) process as

$$\boldsymbol{X}_t = \boldsymbol{\mu} + \sum_{i=0}^{\infty} \boldsymbol{\Phi}_1^i \boldsymbol{Z}_{t-i},$$

where the mean vector of $\boldsymbol{X}_t$, $E(\boldsymbol{X}_t) = \boldsymbol{\mu} = \boldsymbol{\nu} + \boldsymbol{\Phi}_1 \boldsymbol{\nu} + \boldsymbol{\Phi}_1^2 \boldsymbol{\nu} + \cdots$ if $\boldsymbol{\Phi}_1$ has all of its eigenvalues less than 1 in absolute value.

Since the condition on the eigenvalues of the matrix $\boldsymbol{\Phi}_1$ is of importance, we call a VAR(1) process *stable* if

$$\det(\boldsymbol{I}_k - z\boldsymbol{\Phi}_1) \neq 0 \quad \text{for } |z| \leq 1.$$

Note that this is equivalent to the statement that the eigenvalues of $\boldsymbol{\Phi}_1$ are less than 1 in modulus since the eigenvalues of $\boldsymbol{\Phi}_1$ are defined to be the values λ satisfying the equation $\det(\boldsymbol{\Phi}_1 - \lambda \boldsymbol{I}_k) = 0$. For a general VAR($p$) process, we can extend the previous discussion by writing any general VAR(p) in a VAR(1) form. Specifically, we can write

$$
\tilde{\boldsymbol{X}}_t = \begin{pmatrix} \boldsymbol{X}_t \\ \boldsymbol{X}_{t-1} \\ \vdots \\ \boldsymbol{X}_{t-p+1} \end{pmatrix}, \quad \tilde{\boldsymbol{\nu}} = \begin{pmatrix} \boldsymbol{\nu} \\ 0 \\ \vdots \\ 0 \end{pmatrix}, \quad \tilde{\boldsymbol{\Phi}} = \begin{pmatrix} \boldsymbol{\Phi}_1 & \boldsymbol{\Phi}_2 & \cdots & \boldsymbol{\Phi}_{p-1} & \boldsymbol{\Phi}_p \\ \boldsymbol{I}_k & 0 & \cdots & 0 & 0 \\ 0 & \boldsymbol{I}_k & \cdots & 0 & 0 \\ \vdots & \vdots & \ddots & \vdots & \vdots \\ 0 & 0 & \cdots & \boldsymbol{I}_k & 0 \end{pmatrix},
$$

and

$$
\tilde{\boldsymbol{Z}}_t = \begin{pmatrix} \boldsymbol{Z}_t \\ 0 \\ \vdots \\ 0 \end{pmatrix},
$$

where $\tilde{\boldsymbol{X}}_t$, $\tilde{\boldsymbol{\nu}}$, and $\tilde{\boldsymbol{Z}}_t$ are of dimensions $kp \times 1$ and $\tilde{\boldsymbol{\Phi}}$ is $kp \times kp$. Then the original VAR(p) model,

$$
\boldsymbol{X}_t = \boldsymbol{\nu} + \boldsymbol{\Phi}_1 \boldsymbol{X}_{t-1} + \cdots + \boldsymbol{\Phi}_p \boldsymbol{X}_{t-p} + \boldsymbol{Z}_t,
$$

can be expressed as

$$
\begin{aligned}
\tilde{\boldsymbol{X}}_t &= \begin{pmatrix} \boldsymbol{X}_t \\ \boldsymbol{X}_{t-1} \\ \vdots \\ \boldsymbol{X}_{t-p+1} \end{pmatrix} \\
&= \begin{pmatrix} \boldsymbol{\nu} \\ 0 \\ \vdots \\ 0 \end{pmatrix} + \begin{pmatrix} \boldsymbol{\Phi}_1 & \boldsymbol{\Phi}_2 & \cdots & \boldsymbol{\Phi}_{p-1} & \boldsymbol{\Phi}_p \\ \boldsymbol{I}_k & 0 & \cdots & 0 & 0 \\ \vdots & \vdots & \ddots & \vdots & \vdots \\ 0 & 0 & \cdots & \boldsymbol{I}_k & 0 \end{pmatrix} \begin{pmatrix} \boldsymbol{X}_{t-1} \\ \boldsymbol{X}_{t-2} \\ \vdots \\ \boldsymbol{X}_{t-p} \end{pmatrix} + \begin{pmatrix} \boldsymbol{Z}_t \\ 0 \\ \vdots \\ 0 \end{pmatrix} \\
&= \tilde{\boldsymbol{\nu}} + \tilde{\boldsymbol{\Phi}} \tilde{\boldsymbol{X}}_{t-1} + \tilde{\boldsymbol{Z}}_t.
\end{aligned}
$$

Following the foregoing discussion, $\tilde{\boldsymbol{X}}_t$ will be *stable* if

$$
\det(\boldsymbol{I}_{kp} - z\tilde{\boldsymbol{\Phi}}) \neq 0 \quad \text{for } |z| \leq 1.
$$

In this case, the mean vector becomes

$$
\boldsymbol{\mu} = E(\tilde{\boldsymbol{X}}_t) = (\boldsymbol{I}_{kp} - \tilde{\boldsymbol{\Phi}})^{-1} \boldsymbol{\nu},
$$

and the autocovariances are

$$\Gamma(h) = \sum_{i=0}^{\infty} \tilde{\Phi}^{i+h} \Sigma_{\tilde{Z}} (\tilde{\Phi}^i)',$$

where $\Sigma_{\tilde{Z}} = E(\tilde{Z}_t \tilde{Z}_t')$. Using the $k \times kp$ matrix

$$J = (I_k, 0, \ldots, 0),$$

the process X_t can be obtained by setting $X_t = J\tilde{X}_t$. Since $\{\tilde{X}_t\}$ is a well-defined stochastic process, so is $\{X_t\}$. Its mean vector is $E(X_t) = J\mu$, which is constant for all t, and its autocovariances $\Gamma_X(h) = J\Gamma_{\tilde{X}}(h)J'$ are also times invariant.

Further, one can show that

$$\det(I_{kp} - z\tilde{\Phi}) = \det(I_k - \Phi_1 z - \cdots - \Phi_p z^p),$$

which is called the *reverse characteristic polynomial* of the VAR(p) process. Hence, the VAR(p) process is stable if its reverse characteristic polynomial has no roots on the complex unit circle. Formally, a VAR(p) process $\{X_t\}$ is said to be stable if

$$\det(I_k - \Phi_1 z - \cdots - \Phi_p z^p) \neq 0 \quad \text{for } |z| \leq 1.$$

This condition is also called the *stability condition* and it is easy to check for a VAR model. Consider, for example, the two-dimensional VAR(2) process

$$X_t = \nu + \begin{pmatrix} 0.5 & 0.1 \\ 0.4 & 0.5 \end{pmatrix} X_{t-1} + \begin{pmatrix} 0 & 0 \\ 0.25 & 0 \end{pmatrix} X_{t-2} + Z_t.$$

Its reverse characteristic polynomial is

$$\det \left[\begin{pmatrix} 1 & 0 \\ 0 & 1 \end{pmatrix} - \begin{pmatrix} 0.5 & 0.1 \\ 0.4 & 0.5 \end{pmatrix} z - \begin{pmatrix} 0 & 0 \\ 0.25 & 0 \end{pmatrix} z^2 \right]$$

$$= 1 - z + 0.21z^2 - 0.025z^3.$$

The roots for this polynomial are

$$z_1 = 1.3, \quad z_2 = 3.55 + 4.26i, \quad \text{and } z_3 = 3.55 - 4.26i.$$

You can find the roots using the SPLUS command

```
> polyroot(c(1,-1,0.21,-0.025))
[1] 1.299957+9.233933e-15i 3.550022+4.262346e+00i
[3] 3.550022-4.262346e+00i
```

Note that the modulus of z_2 and z_3 is $|z_2| = |z_3| = \sqrt{3.55^2 + 4.26^2} = 5.545$. Thus the process satisfies the stability condition since all roots are outside the unit circle. A common feature of stable VAR processes (when you plot them with, say, `tsplot` in SPLUS) is that they fluctuate around constant means, and their variability does not change as they wander along.

10.5 EXAMPLE OF INFERENCES FOR VAR

Most of the inferences on estimation and testing for VAR models are similar to those of the univariate case, although notations for writing down the estimator can become notoriously complicated. Again, we shall not discuss these details but mention that most of the estimation methods are still likelihood based. As in the univariate case, computations of likelihood procedures can be very tricky when MA parts are incorporated, and this problem becomes more severe for VARMA models. Notice that at the writing of this book, SPLUS only supports VAR models. Although this is somewhat limited, it should be pointed out that VAR models work quite well in many of the financial and econometric applications. Consequently, the rest of this section is devoted to an example to illustrate the VAR features of SPLUS.

Consider a three-dimensional system consisting of quarterly seasonally adjusted German fixed investments disposable income, and consumption expenditures in billions of deutsche marks. The data are from 1960 to 1982 but we will only use data up to 1978 for the analysis. This data set can be found in Lütkepohl (1993) and is also stored on the Web page for this book under the file name `weg.dat`. The actual data file consists of five columns. The first column contains the year, the second the quarter, and columns 3 through 5 contain the actual data.

We first read the data in SPLUS using the `read.table` command, and let SPLUS know that we are dealing with time series data using the `rts` command as follows:

```
> weg_read.table('weg.dat',row.names=NULL)
> weg
> weg
      V1 V2  V3   V4    V5
 1 1960  1 180  451   415
 2 1960  2 179  465   421
 3 1960  3 185  485   434
 4 1960  4 192  493   448
 5 1961  1 211  509   459
 6 1961  2 202  520   458
 .    .  .   .    .     .
 .    .  .   .    .     .
75 1978  3 675 2121  1831
76 1978  4 700 2132  1842
77 1979  1 692 2199  1890
78 1979  2 759 2253  1958
79 1979  3 782 2276  1948
80 1979  4 816 2318  1994
81 1980  1 844 2369  2061
82 1980  2 830 2423  2056
```

```
83 1980   3 853 2457 2102
84 1980   4 852 2470 2121
85 1981   1 833 2521 2145
86 1981   2 860 2545 2164
87 1981   3 870 2580 2206
88 1981   4 830 2620 2225
89 1982   1 801 2639 2235
90 1982   2 824 2618 2237
91 1982   3 831 2628 2250
92 1982   4 830 2651 2271
> WEG_weg[1:76,]
> invest_rts(WEG$V3,start=c(1960,1),freq=4)
> income_rts(WEG$V4,start=c(1960,1),freq=4)
> consum_rts(WEG$V5,start=c(1960,1),freq=4)
> invest
         1   2   3   4
1960: 180 179 185 192
1961: 211 202 207 214
1962: 231 229 234 237
1963: 206 250 259 263
1964: 264 280 282 292
1965: 286 302 304 307
1966: 317 314 306 304
1967: 292 275 273 301
1968: 280 289 303 322
1969: 315 339 364 371
1970: 375 432 453 460
1971: 475 496 494 498
1972: 526 519 516 531
1973: 573 551 538 532
1974: 558 524 525 519
1975: 526 510 519 538
1976: 549 570 559 584
1977: 611 597 603 619
1978: 635 658 675 700
 start deltat frequency
  1960   0.25         4
```

We plot the series in Figure 10.1 and their autocorrelations in Figure 10.2. Notice that in Figure 10.2, the diagonal elements are the ACFs of the univariate series of each component, and the superdiagonal plots are the cross ACFs between the components. For the subdiagonal plots, observe that the lags are negative. These are the plots of $R_{ij}(-h)$, which according to part 1 of Theorem 10.1 equal $R_{ji}(h)$. This provides a way to get hold of $R_{ji}(h)$.

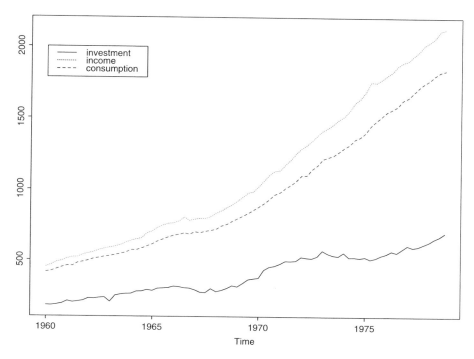

Fig. 10.1   German investment, income, and consumption data.

For example, the (2,1) entry of Figure 10.2 represents $R_{12}(-h) = R_{21}(h)$. Therefore, $R_{21}(13) = R_{12}(-13) = 0.058$.

```
> weg.ts_ts.union(invest,income,consum)
> tsplot(weg.ts)
> legend(1960,2000,legend=c("income","consumption",
+ "investment"),lty=1:3)
> acf(weg.ts)
```

The original data have a trend and are thus considered nonstationary. The trend is removed by taking first differences of logarithms:

```
> dlinv1_diff(log(invest))
> dlinc1_diff(log(income))
> dlcon1_diff(log(consum))
> par(mfrow=c(3,1))
> tsplot(dlinv1);title("investment")
> tsplot(dlinc1);title("income")
> tsplot(dlcon1);title("consumption")
> dweg.ts_ts.union(dlinv1,dlinc1,dlcon1)
> acf(dweg.ts)
```

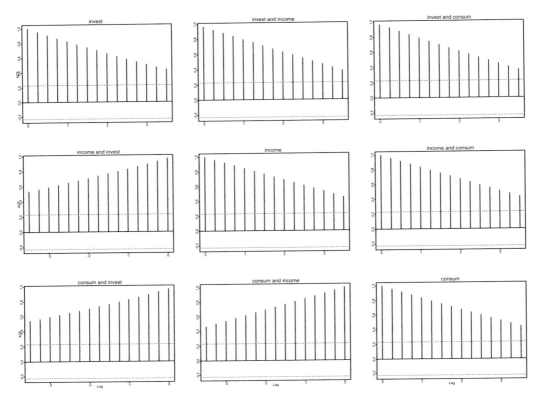

Fig. 10.2 Sample ACF of German data.

The time series plots of the first differences as well as the sample autocorrelation functions are shown in Figures 10.3 and 10.4, respectively.

We fit a VAR model using the **ar** function, which uses the AIC criterion to select the order.

```
> weg.ar_ar(dweg.ts)
> weg.ar$aic
 [1] 12.536499  7.375732  0.000000 12.605591 17.017822
 [6] 30.794067 35.391541 50.271301 54.909058 66.699646
[11] 71.657837 80.051147 92.501953 93.377502
> weg.ar$order
[1] 2
> weg.ar$ar[1,,]
              [,1]        [,2]        [,3]
[1,] -0.309421629  0.1552022   0.8746032
[2,]  0.041505784 -0.1069321   0.2424691
[3,] -0.003082463  0.2385896  -0.2722486
```

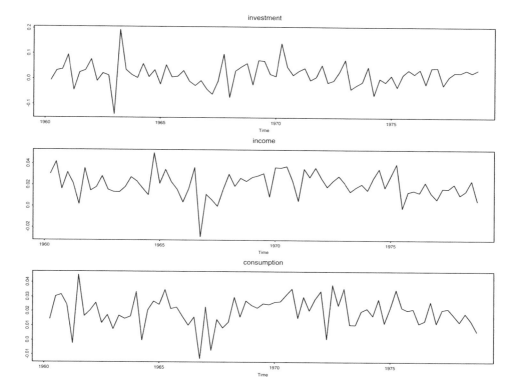

Fig. 10.3 First differences of logarithms of German investment, income, and consumption.

```
> weg.ar$ar[2,,]
            [,1]        [,2]          [,3]
[1,] -0.15150616 0.14152186  0.83874440
[2,]  0.04778249 0.03499804 -0.02906964
[3,]  0.03408687 0.35225770 -0.03116566
> weg.ar$var.pred
            [,1]          [,2]          [,3]
[1,] 2.156003e-03 7.242693e-05 1.267914e-04
[2,] 7.242693e-05 1.486544e-04 6.316167e-05
[3,] 1.267914e-04 6.316167e-05 9.070559e-05
```

This means that the parameter estimates are

$$\hat{\mathbf{\Phi}}_1 = \begin{pmatrix} -0.309 & 0.155 & 0.875 \\ 0.042 & -0.107 & 0.242 \\ -0.003 & 0.239 & -0.272 \end{pmatrix}, \quad \hat{\mathbf{\Phi}}_2 = \begin{pmatrix} -0.152 & 0.142 & 0.839 \\ 0.048 & 0.035 & -0.030 \\ 0.034 & 0.352 & -0.031 \end{pmatrix},$$

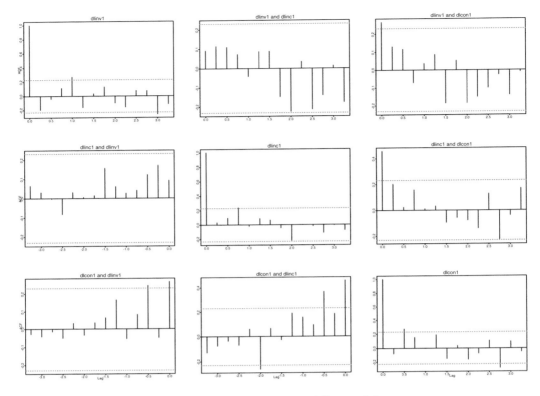

Fig. 10.4 Sample ACF of the differenced data.

and
$$\hat{\mathbf{\Sigma}}_z = \begin{pmatrix} 2.156003e\text{-}03 & 7.242693e\text{-}05 & 1.267914e\text{-}04 \\ 7.242693e\text{-}05 & 1.486544e\text{-}04 & 6.316167e\text{-}05 \\ 1.267914e\text{-}04 & 6.316167e\text{-}05 & 9.070559e\text{-}05 \end{pmatrix}.$$

The residuals can be seen in Figure 10.5.

```
> tsplot(weg.ar$resid)
> legend(1975,-.10,legend=c("income","consumption","investment"),
+ lty=1:3)
```

Finally, using the fitted model for the data from 1960 to 1978, we can attempt to predict the next eight quarters using the function pred.ar, which is presented in the SPLUS help window for ar. Figure 10.6 shows the predicted values (solid) and the actual values (dotted lines).

```
# function to predict using an ar model:
        # ahead gives the number of predictions to make

        pred.ar <- function(series, ar.est, ahead = 1)
```

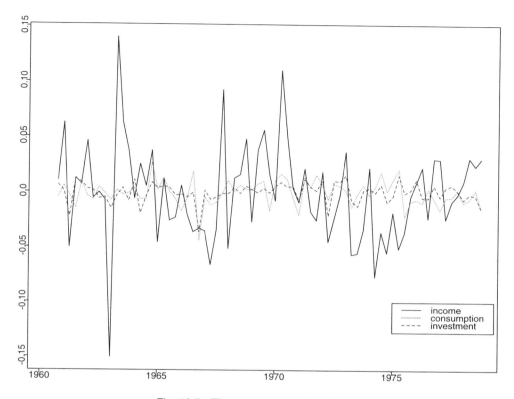

Fig. 10.5   Time series plot of residuals.

```
{
        order <- ar.est$order
        series <- as.matrix(series)
        pred.out <- array(NA, dim = c(order + ahead,
                ncol(series)),dimnames = list(NULL,
                dimnames(series)[[2]]))
        mean.ser <- apply(series, 2, mean)
        ser.cent <- sweep(series, 2, mean.ser)
        pred.out[seq(order),  ] <- ser.cent[rev(nrow(
                series) - seq(order) + 1),  ]
        for(i in (order + 1):nrow(pred.out)) {
                pred.out[i,  ] <- apply(aperm(ar.est$ar,
                        c(1, 3, 2)) * as.vector(pred.out
                        [i - seq(order),  ]), 3, sum)
        }
        sweep(pred.out[ - seq(order),  , drop = F], 2,
                mean.ser, "+")
}
```

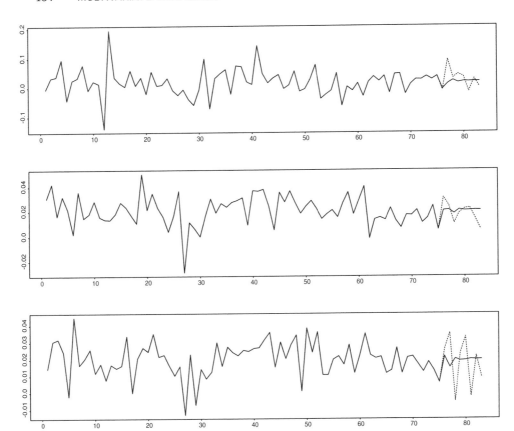

Fig. 10.6 Predicted values for German series.

```
> weg.pred_pred.ar(dweg.ts,weg.ar,ahead=8)
> weg.pred
           dlinv1       dlinc1      dlcon1
[1,] -0.008301602 0.02043956 0.02162485
[2,]  0.011163449 0.02080164 0.01499163
[3,]  0.021437457 0.01790787 0.02019181
[4,]  0.013896021 0.02104020 0.01905175
[5,]  0.018114067 0.02035429 0.01930151
[6,]  0.017550547 0.02044569 0.01993871
[7,]  0.017769750 0.02073731 0.01968315
[8,]  0.018156437 0.02061101 0.01981476
```

10.6 EXERCISES

1. Consider the German economy example discussed in Section 10.5.

   (a) Fit a univariate time series to each component in this data set.

   (b) Combine the three fits that you obtained from part (a) and form a three-dimensional multivariate time series by stacking them together. How do you compare this model with the VAR(2) used in Section 10.5?

   (c) Perform a forecast based on your combined series and compare your result with those given in Section 10.5.

2. (a) For any matrix Φ such that $\Phi^2 = 0$, show that $(I - \Phi B)^{-1} = (I + \Phi B)$, where B is the backshift operator.

   (b) Establish the reverse characteristic polynomial result for the case $k = 2$ and $p = 2$; that is, for a VAR(2) model $(I_2 - \Phi_1 B - \Phi_2 B^2)X_t = Z_t$, show that

$$\det(I_2 - \Phi_1 z - \Phi_2 z^2) = \det(I_4 - z\tilde{\Phi}),$$

   where

$$\tilde{\Phi} = \begin{pmatrix} \Phi_1 & \Phi_2 \\ I_2 & 0 \end{pmatrix}.$$

3. In the United States of Wonderland, the growth rates for income (GNP) and money demand (M2) and an interest rate (IR) are related in the following VAR(2) model:

$$\begin{pmatrix} GNP_t \\ M2_t \\ IR_t \end{pmatrix} = \begin{pmatrix} 2 \\ 1 \\ 0 \end{pmatrix} + \begin{pmatrix} 0.7 & 0.1 & 0 \\ 0 & 0.4 & 0.1 \\ 0.9 & 0 & 0.8 \end{pmatrix} \begin{pmatrix} GNP_{t-1} \\ M2_{t-1} \\ IR_{t-1} \end{pmatrix}$$
$$+ \begin{pmatrix} -0.2 & 0 & 0 \\ 0 & 0.1 & 0.1 \\ 0 & 0 & 0 \end{pmatrix} \begin{pmatrix} GNP_{t-2} \\ M2_{t-2} \\ IR_{t-2} \end{pmatrix} + \begin{pmatrix} Z_{1t} \\ Z_{2t} \\ Z_{3t} \end{pmatrix},$$

   where

$$\Sigma_Z = \begin{pmatrix} 0.26 & 0.03 & 0 \\ 0.03 & 0.09 & 0 \\ 0 & 0 & 0.81 \end{pmatrix} = PP', \quad P = \begin{pmatrix} 0.5 & 0.1 & 0 \\ 0 & 0.3 & 0 \\ 0 & 0 & 0.9 \end{pmatrix}.$$

   (a) Show that the process $X_t = (GNP_t, M2_t, IR_t)'$ is stable.

   (b) Determine the mean vector of X_t.

   (c) Write the process of X_t in the X_t, VAR(1) form.

11

State Space Models

11.1 INTRODUCTION

State space representations of time series have had a long history. They have found applications in diverse disciplines. Under a state space setting, an extremely rich class of time series, including and going well beyond the linear models considered in this book, can be formulated as special cases of the general state space model defined below. Early accounts of state space models and the associated Kalman filter recursions can be found in control engineering literature [see, e.g., Kailath (1980) or Hannan and Deistler (1988)]. A recent account of time series analysis that emphasizes the state space methodology is given in Shumway and Stoffer (2000). Other references for state space frameworks in the econometric literature are Harvey (1993) and Aoki (1990). A very general Bayesian state space framework under the context of the dynamic linear model is given in West and Harrison (1997).

11.2 STATE SPACE REPRESENTATION

A state space model of a given time series (possibly multivariate) $\{\boldsymbol{Y}_t : t = 1, 2, \dots\}$ consists of two equations: the observation equation and the state equation. The observation equation relates the observed data $\{\boldsymbol{Y}_t\}$ to the underlying states $\{\boldsymbol{X}_t : t = 1, 2, \dots\}$ (most likely unobservable or latent) that govern the system via

$$\boldsymbol{Y}_t = G_t \boldsymbol{X}_t + \boldsymbol{W}_t, \qquad (11.1)$$

where $\{\boldsymbol{W}_t\} \sim \mathrm{WN}(\boldsymbol{0}, R_t)$ represents the observation errors and G_t is a sequence of matrices. The state equation describes the evolution of the states via

$$\boldsymbol{X}_{t+1} = F_t \boldsymbol{X}_t + \boldsymbol{V}_t, \tag{11.2}$$

where $\{F_t\}$ is a sequence of matrices and $\{\boldsymbol{V}_t\} \sim \mathrm{WN}(\boldsymbol{0}, Q_t)$ denotes the errors incurred in describing the states. It is usually assumed that the observation errors and the state errors are uncorrelated [i.e., $E(\boldsymbol{W}_s \boldsymbol{V}_t) = \boldsymbol{0}$ for all s and t]. To complete the specification, the initial value of $\boldsymbol{X}_1$ is assumed to be uncorrelated with all the errors $\{\boldsymbol{V}_t\}$ and $\{\boldsymbol{W}_t\}$.

Definition 11.1 *A time series $\{\boldsymbol{Y}_t\}$ has a state space representation if there exists a state space model for $\{\boldsymbol{Y}_t\}$ as specified in equations* (11.1) *and* (11.2).

Remarks

1. The uncorrelatedness assumption between the errors can be relaxed. Further, one can include a control term $H_t \boldsymbol{u}_t$ in the state equation (11.2) to account for exogenous information. In an ARMA context, such an inclusion is usually known as an *ARMAX model*.

2. In many applications, the observation and system matrices $\{F_t\}, \{G_t\}$, $\{R_t\}, \{Q_t\}$ are independent of time, resulting in a time-invariant system. We deal only with time-invariant systems in this book.

3. By iterating the state equation, it can easily be seen that

$$\boldsymbol{X}_t = f_t(\boldsymbol{X}_1, \boldsymbol{V}_1, \dots, \boldsymbol{V}_{t-1}),$$
$$\boldsymbol{Y}_t = g_t(\boldsymbol{X}_1, \boldsymbol{V}_1, \dots, \boldsymbol{V}_{t-1}, \boldsymbol{W}_t),$$

for some functions f_t and g_t. In particular, we can derive

$$E(\boldsymbol{V}_t \boldsymbol{X}'_s) = \boldsymbol{0}, \qquad E(\boldsymbol{V}_t \boldsymbol{Y}'_s) = \boldsymbol{0}, \ 1 \le s \le t,$$
$$E(\boldsymbol{W}_t \boldsymbol{X}'_s) = \boldsymbol{0}, \ 1 \le s \le t, \text{ and } E(\boldsymbol{W}_t \boldsymbol{Y}'_s) = \boldsymbol{0}, \ 1 \le s < t.$$

The usefulness of a state space representation is that it is highly flexible and can be used to represent a large number of time series models. As given, neither $\{\boldsymbol{X}_t\}$ or $\{\boldsymbol{Y}_t\}$ is necessarily stationary. Whenever a simple state space representation can be found, we can study the behavior of the states $\{\boldsymbol{X}_t\}$ from the observations $\{\boldsymbol{Y}_t\}$ via the observation equation (11.1). Notice that the states and the observations do not have to be completely unrelated. Past observations can be components in a state. To illustrate this point, consider an $\mathrm{AR}(p)$ model.

Example 11.1 *Let $\{Y_t\}$ be a causal $\mathrm{AR}(p)$ model $\phi(B)Y_t = Z_t$. To express it in a state space representation, define $\boldsymbol{X}_t = (Y_{t-p+1}, Y_{t-p+2}, \dots, Y_t)'$ and*

$\boldsymbol{Z}_{t+1} = (0, \dots, 0, Z_{t+1})'$. Then

$$\boldsymbol{X}_{t+1} = \begin{pmatrix} 0 & 1 & 0 & \cdots & 0 \\ 0 & 0 & 1 & \cdots & 0 \\ \vdots & \vdots & \vdots & \ddots & \vdots \\ 0 & 0 & 0 & \cdots & 1 \\ \phi_p & \phi_{p-1} & \phi_{p-2} & \cdots & \phi_1 \end{pmatrix} \boldsymbol{X}_t + \boldsymbol{Z}_{t+1}$$

$$:= F\boldsymbol{X}_t + \boldsymbol{Z}_{t+1},$$

$$Y_t = (0, \cdots, 0, 1)\boldsymbol{X}_t.$$

These equations have the required forms of (11.1) *and* (11.2). *In this case,* $\boldsymbol{W}_t = \boldsymbol{0}$ *and* $\boldsymbol{V}_t = \boldsymbol{Z}_{t+1}$. *The causality condition is equivalent to the condition that the state equation is stable (i.e., the eigenvalues of the state matrix* F *all lie inside the unit disk).*   □

As a second example, let us consider the *structural model*, which permits random variation in the trend process. Recall that in Chapter 1, we decomposed a time series into $Y_t = M_t + W_t$, where M_t represents the trend.

Example 11.2 *Consider the random walk plus noise model represented by*

$$Y_t = M_t + W_t, W_t \sim \mathrm{WN}(0, \sigma_w^2),$$

$$M_{t+1} = M_t + V_t, V_t \sim \mathrm{WN}(0, \sigma_v^2),$$

with the initial value $M_1 = m_1$ *fixed.*

Notice that this model is already written in a state space representation. The trend component is unobservable and follows a random walk model, while the observation is expressed in a signal plus noise equation. This is sometimes known as a local level *or* random walk plus noise model. *The signal-to-noise ratio is defined as*

$$\mathrm{SNR} = \frac{\sigma_v^2}{\sigma_w^2},$$

which is an important factor in determining the features of the model. The larger this factor is, the more information can be gathered about the signal M_t. *On the other hand, when* σ_v^2 *is zero,* M_t *is constant and the model reduces to a trivial constant mean model. If we extend the trend equation to incorporate a linear trend, we have*

$$M_t = M_{t-1} + B_{t-1} + V_{t-1},$$

$$B_t = B_{t-1} + U_{t-1}, \quad U_t \sim \mathrm{WN}(0, \sigma_u^2),$$

where M_t *represents a local linear trend with slope* B_{t-1} *at time* $t - 1$. *To write this equation in a state space form, define* $\boldsymbol{X}_t = (M_t, B_t)'$. *Then*

$$\boldsymbol{X}_{t+1} = \begin{pmatrix} 1 & 1 \\ 0 & 1 \end{pmatrix} \boldsymbol{X}_t + \boldsymbol{V}_t,$$

where $\boldsymbol{V}_t = (V_t, U_t)'$. This constitutes the state equation. For the observation equation, we have

$$Y_t = (1, 0)\boldsymbol{X}_t + W_t,$$

where we assume $\{\boldsymbol{X}_1, U_1, V_1, W_1, U_2, V_2, W_2, \dots\}$ to be an uncorrelated sequence of random variables. In this model,

$$F = \begin{pmatrix} 1 & 1 \\ 0 & 1 \end{pmatrix}, \ G = (1, 0), \ Q = \begin{pmatrix} \sigma_v^2 & 0 \\ 0 & \sigma_u^2 \end{pmatrix}, \ \text{and } R = \sigma_w^2. \qquad \square$$

These two examples demonstrate the versatility of the state space formulation. In general, we can write an ARIMA(p, d, q) model or any randomly varying trend plus noise component models in state space forms, although finding one that is convenient to work with could be a tricky exercise.

11.3 KALMAN RECURSIONS

The fundamental problems associated with a state space model can be collectively classified into one of the following three categories, which are concerned with estimating the state vector $\boldsymbol{X}_t$ in terms of the observations $\boldsymbol{Y}_1, \boldsymbol{Y}_2, \dots$ and an initial value $\boldsymbol{Y}_0$. Estimation of $\boldsymbol{X}_t$ in terms of

1. $\boldsymbol{Y}_0, \dots, \boldsymbol{Y}_{t-1}$ defines a Kalman prediction problem.

2. $\boldsymbol{Y}_0, \dots, \boldsymbol{Y}_t$ defines a Kalman filtering problem.

3. $\boldsymbol{Y}_0, \dots, \boldsymbol{Y}_n (n > t)$ defines a Kalman smoothing problem.

Each of these problems can be solved by using an appropriate set of Kalman recursions. Usually, $\boldsymbol{Y}_0$ is chosen to be equal to the vector $(1, \dots, 1)'$. Define $P_t(X_i) = P(X_i | \boldsymbol{Y}_0, \dots, \boldsymbol{Y}_t) = P_{\bar{sp}\{\boldsymbol{Y}_0, \dots, \boldsymbol{Y}_t\}} X_i, \ i = 1, \dots, v$ as the best linear predictor of X_i in terms of $\boldsymbol{Y}_0, \dots, \boldsymbol{Y}_t$. Further, introduce the notation

$$P_t(\boldsymbol{X}) = (P_t(X_1), \dots, P_t(X_v))',$$

$$\hat{\boldsymbol{X}}_t = P_{t-1}(\boldsymbol{X}_t), \ \text{the one-step-ahead prediction,}$$

$$\Omega_t = E[(\boldsymbol{X}_t - \hat{\boldsymbol{X}}_t)(\boldsymbol{X}_t - \hat{\boldsymbol{X}}_t)'],$$

where Ω_t denotes the one-step-ahead prediction error covariance matrix.

Theorem 11.1 (Kalman Prediction) *For a given state space model, the one-step-ahead predictors and their error covariance matrices are determined uniquely by the initial conditions*

$$\hat{\boldsymbol{X}}_1 = P(\boldsymbol{X}_1 | \boldsymbol{Y}_0), \ \ \Omega_1 = E[(\boldsymbol{X}_1 - \hat{\boldsymbol{X}}_1)(\boldsymbol{X}_1 - \hat{\boldsymbol{X}}_1)'],$$

and the recursions for $t = 1, 2, \dots$:

$$\hat{\boldsymbol{X}}_{t+1} = F_t \hat{\boldsymbol{X}}_t + \Theta_t \Delta_t^{-1} (\boldsymbol{Y}_t - G_t \hat{\boldsymbol{X}}_t),$$

$$\Omega_{t+1} = F_t \Omega_t F_t' + Q_t - \Theta_t \Delta_t^{-1} \Theta_t',$$

where

$$\Delta_t = G_t \Omega_t G'_t + R_t, \quad \Theta_t = F_t \Omega_t G'_t,$$

and Δ_t^{-1} denotes the generalized inverse of the matrix Δ_t.

Theorem 11.2 (Kalman Filtering) *The filtered estimates $X_{t|t} = P_t(X_t)$ and their error covariance matrices $\Omega_{t|t} = E[(X_t - \hat{X}_{t|t})(X_t - \hat{X}_{t|t})']$ are determined by the relations*

$$P_t(X_t) = P_{t-1}X_t + \Omega_t G'_t \Delta_t^{-1}(Y_t - G_t \hat{X}_t)$$

and

$$\Omega_{t|t} = \Omega_t - \Omega_t G'_t \Delta_t^{-1} G_t \Omega'_t.$$

Theorem 11.3 (Kalman Smoothing) *The smoothed estimates $X_{t|n} = P_n(X_t)$ and the error covariance matrices $\Omega_{t|n} = E[(X_t - X_{t|n})(X_t - X_{t|n})']$ are determined for fixed t by the following recursions, which can be solved successively for $n = t, t+1, \ldots$:*

$$P_n(X_t) = P_{n-1}(X_t) + \Omega_{t,n} G'_n \Delta_n^{-1}(Y_n - G_n \hat{X}_n),$$
$$\Omega_{t,n+1} = \Omega_{t,n}[F_n - \Theta_n \Delta_n^{-1} G_n]',$$
$$\Omega_{t|n} = \Omega_{t|n-1} - \Omega_{t,n} G'_n \Delta_n^{-1} G_n \Omega'_{t,n},$$

where $\Omega_{t,n} = E[(X_t - \hat{X}_t)(X_n - \hat{X}_n)']$ with initial conditions $P_{t-1}(X_t) = \hat{X}_t$ and $\Omega_{t,t} = \Omega_{t|t-1} = \Omega_t$ being determined from the Kalman prediction recursions.

Proofs of these theorems can be found in Brockwell and Davis (1991). Although tedious, these results provide the necessary algorithms to compute prediction in an on-line manner, and they have been programmed in many packages, including SPLUS. Furthermore, by combining these algorithms with Gaussian likelihood, we can compute the MLE very efficiently. Consider the problem of finding the parameter θ that maximizes the likelihood function for given $Y_1, \ldots, Y_n$. Let the conditional density of Y_t given $(Y_{t-1}, \ldots, Y_0)$ be $f_t(Y_t | Y_{t-1}, \ldots, Y_0)$. The likelihood function of $Y_1, \ldots, Y_n$ conditional on Y_0 can be written as

$$L(\theta, Y_1, \ldots, Y_n) = \prod_{t=1}^{n} f_t(Y_t | Y_{t-1}, \ldots, Y_1).$$

Assuming all the errors to be jointly Gaussian and letting $I_t = Y_t - P_{t-1}Y_t = Y_t - G\hat{X}_t$ as the one-step-ahead prediction error with covariance matrix $\Delta_t = E(I_t I'_t)$, the conditional density can be written as

$$f_t(Y_t | Y_{t-1}, \ldots, Y_1) = (2\pi)^{-w/2}(\det \Delta_t)^{-1/2} \exp\left(-\frac{1}{2} I'_t \Delta_t^{-1} I_t\right).$$

Therefore, the likelihood function of the observations $\boldsymbol{Y}_1, \ldots, \boldsymbol{Y}_n$ is given by

$$L(\boldsymbol{\theta}, \boldsymbol{Y}_1, \ldots, \boldsymbol{Y}_n) = (2\pi)^{-nw/2} \left(\prod_{j=1}^{n} \Delta_j \right)^{-1/2} \exp\left(-\frac{1}{2} \sum_{j=1}^{n} \boldsymbol{I}'_j \Delta_j^{-1} \boldsymbol{I}_j \right).$$

In particular, when $w = 1$ (i.e., the series is univariate), we have

$$L(\boldsymbol{\theta}, Y_1, \ldots, Y_n) = (2\pi)^{-n/2} \left(\prod_{j=0}^{n-1} v_j \right)^{-1/2} \exp\left[-\frac{1}{2} \sum_{j=1}^{n} (Y_j - \hat{Y}_j)^2 / v_{j-1} \right],$$

where $\hat{Y}_j = P_{j-1}(Y_j)$ denotes the one-step-ahead prediction of Y_j with $\hat{Y}_1 = 0$ and $v_j = E(Y_{j+1} - \hat{Y}_{j+1})^2$ denotes the variance of the one-step-ahead prediction (innovation) error. Using Kalman filtering, both $\hat{Y}_j$ and v_j can be updated recursively. This algorithm has been used in SPLUS to evaluate the MLE for an ARMA model.

In general, given the observations, an initial value $\boldsymbol{Y}_0$, and a starting value $\boldsymbol{\theta}_0$, the likelihood function can be maximized numerically from the preceding equation with the aid of the Kalman recursions. Furthermore, once the MLE is found, we can compute forecasts based on the state space representation and mean square errors by means of Kalman predictions.

11.4   STOCHASTIC VOLATILITY MODELS

One of the applications of a state space representation in finance is in modeling heteroskedasticity. In addition to GARCH models, stochastic volatility models offer a useful alternative for describing volatility clustering. It is not our intention to provide a comprehensive account of the developments of stochastic volatility models here; interested readers may find detailed discussions about various aspects of stochastic volatility models in Ghysels, Harvey, and Renault (1996) and Taylor (1994). Rather, our aim here is to introduce the stochastic volatility model through the state space representation.

In stochastic volatility models, the instantaneous variance of the series observed is modeled as a nonobservable or latent process. Let $\{x_t\}$ denote the returns on an equity. A basic setup of a stochastic volatility model takes the form

$$\begin{cases} x_t = \sigma_t \xi_t, \\ \sigma_t = \exp(v_t/2), \end{cases} \tag{11.3}$$

where $\{\xi_t\}$ is usually assumed to be a sequence of independent standard normal random variables, and the log volatility sequence $\{v_t\}$ satisfies an ARMA relation

$$\phi(B)v_t = \theta(B)\eta_t. \tag{11.4}$$

Here, $\{\eta_t\}$ is a Gaussian white noise sequence with variance τ, $\phi(\cdot)$ and $\theta(\cdot)$ are polynomials of order p, q, respectively, with all their roots outside the unit circle and with no common root, and B is the backshift operator $Bx_t = x_{t-1}$. Conceptually, this represents an extension with respect to GARCH models, since the evolution of the volatility is not determined completely by the past observations. It also includes a stochastic component and allows for a more flexible mechanism. Unfortunately, since $\{\sigma_t\}$ is not observable, the method of quasi maximum likelihood cannot be directly applicable. By letting $y_t = \log x_t^2$, $u_t = \log Z_t^2$, and taking the log and squaring equation (11.3), we have

$$y_t = v_t + u_t, \tag{11.5}$$

$$\phi(B)v_t = \theta(B)\eta_t. \tag{11.6}$$

In this expression, the log volatility sequence satisfies a linear state space model with state equation (11.6) and observation equation (11.5), while the original process $\{\sigma_t\}$ follows a nonlinear state space model. To complicate matters further, the observation error $u_t = \log \xi_t^2$ in (11.5) is non-Gaussian. Consequently, direct applications of the Kalman filter method for linear Gaussian state space models seem unrealistic. Several estimation procedures have been developed for SV models to circumvent some of these difficulties. Melino and Turnbull (1990) use a generalized method of moments (GMM), which is straightforward to implement but not efficient. Harvey, Ruiz, and Shephard (1994) propose a quasi maximum likelihood approach based on approximating the observation error $\{u_t\}$ by a mixture of Gaussian random variables which renders (11.5) and (11.6) into a linear Gaussian state space setup. A Bayesian approach is taken by Jacquier, Polson, and Rossi (1994). Kim, Shephard, and Chib (1998) suggest a simulation-based exact maximum likelihood estimator, and Sandmann and Koopman (1998) propose a Monte Carlo maximum likelihood procedure. Although each of these methods is reported to work well under certain conditions, it is difficult to assess their overall performances across different data sets. Alternatively, the SV model can be considered as a discrete-time realization of a continuous-time process.

In summary, although stochastic volatility models have natural links to state space representations, they are nonlinear and non-Gaussian state space forms. As such, it takes more work to estimate and test for an SV model. Further developments about estimating SV models and long-memory effects can be found in Chan and Petris (2000). From a pure modeling perspective, an SV model seems to be more flexible, but the lack of available softwares for SV models also makes it less readily applicable in practice. Empirical evidence shows that both GARCH and SV models perform similarly, and it is not clear that one form can be uniformly better than the other. Perhaps it is for this reason that GARCH models have been receiving considerably more attention at the user's end.

11.5   EXAMPLE OF KALMAN FILTERING OF TERM STRUCTURE

In this example we illustrate how to use Kalman filters in estimating the simple Vasicek term-structure model discussed in Babbs and Nowman (1999). In the basic framework, consider the instantaneous spot rate $r(t)$ described by

$$r(t) = \mu(t) - \sum_{j=1}^{J} X_j(t),$$

$$dX_j(t) = -\xi_j X_j(t)\,dt + \sigma_j dW_j(t).$$

For a fixed j, the discretized version of the second equation can be written as

$$X_k = \left(1 - \frac{\xi}{n}\right) X_{k-1} + \eta_k$$

$$\sim e^{-\xi/n} X_{k-1} + \eta_k,$$

where $\eta_k \sim N(0, V^2)$. In this setting, $X_1, \ldots, X_J$ represents the current effects of J streams of economic "news" whose impact is described by the state equation. Here, we simplify the case where W_j denotes independent Brownian motions, although this assumption can be relaxed to incorporate correlations among different components as studied in Babbs and Nowman. From now on, we further restrict our attention to a one-factor model, $J = 1$, with constant parameters [i.e., $\mu(t) = \mu, \sigma_j = \sigma, \xi_j = \xi$, and $W_j = W$]. If we denote the market price of risk associated with W as θ, the resulting pricing formula for $B(M, t)$, the price of a unit nominal pure discount bond maturing at time M, is given in equation (5) of Babbs and Nowman. Using the fact that the theoretical interest rate

$$R(M, t) = -\frac{1}{\tau} \log B(t + \tau, t),$$

where $\tau = M - t$ denotes the residual term to maturity, and equations (5), (14), and (15) of Babbs and Nowman, we can write

$$R(M, t) = A_0(\tau) - A_1(\tau)X(t),$$

where

$$A_0(\tau) = R(\infty) - w(\tau), \tag{11.7}$$

$$A_1(\tau) = H(\xi\tau), \tag{11.8}$$

$$R(\infty) = \mu + \theta v - v^2/2,$$

$$w(\tau) = H(\xi\tau)\left(\theta v - \frac{v^2}{2}\right) + \frac{1}{2}H(2\xi\tau)v^2,$$

$$v = \frac{\sigma}{\xi},$$

$$H(x) = \frac{1 - e^{-x}}{x}.$$

When we discretized the equation for $r(t)$ and $X(t)$, a state space formulation resulted, with the following observation and system equations. For each $i = 1, \ldots, N$,

$$R_{ik} = A_0(\tau_i) - A_1(\tau_i)X_k + \epsilon_{ik},$$
$$X_k = e^{-\xi(t_k - t_{k-1})}X_{k-1} + \eta_k,$$

where $\epsilon_{ik} \sim N(0, \sigma_i^2(\psi))$ and $\eta_k \sim N(0, V^2(\psi))$ denote the observation and system noise, respectively. In this setting, the unknown parameters are described collectively by a hyperparameter $\psi = (\mu, \theta, \xi, \sigma)'$, so that the variances of both of the observation and system errors are functions of ψ. Also, note that the state X is unobservable and that the only observable component is the interest rate R_{ik}, where $k = 1, \ldots, n$ denotes the time of the observations and $i = 1, \ldots, N$ denotes the dimension of the observation vector $R_k = (R_{1k}, \ldots, R_{Nk})'$. This framework provides a good candidate for making use of the Kalman filter algorithm to estimate the parameters $\psi = (\mu, \theta, \xi, \sigma)'$.

Specifically, consider the following spot rates example (see

http://economics.sbs.ohio-state.edu/jhm/ts/mcckwon/mccull.htm

for a detailed description of the data). The data are spot interest rates for eight maturities for each month from August 1985 to February 1991 (so that $n = 67$ months in total). These are basically points on the zero-coupon yield curve. The eight maturities chosen are for three and six months; one, two, three, five, seven, and ten years; and $N = 8$. The data are stored in the file zeros.dat on the Web page for this book. In this example, $n = 67$ and $N = 8$, with R_{1k} denoting the three-month rates for $k = 1, \ldots, 67$, R_{2k} denoting the six-month rates for $k = 1, \ldots, 67$, $\ldots$, and R_{8k} denoting the ten-year rates for $k = 1, \ldots, 67$. Writing it in a vector form, the observation equation becomes

$$R_k = d(\psi) + Z(\psi)X_k + \epsilon_k,$$

where $d(\psi) = (A_0(\tau_1), \ldots, A_0(\tau_8))'$ and $Z(\psi) = (A_1(\tau_1), \ldots, A_1(\tau_8))'$ are 8×1 vectors with A_0 and A_1 defined in equations (11.7) and (11.8), respectively, and $\tau_1, \ldots, \tau_8$ denote the residual terms to each maturity. In this equation it is assumed that $\epsilon_k \sim N(0, \Sigma(\psi))$. The state equation is the same as before:

$$X_k = e^{-\xi(t_k - t_{k-1})}X_{k-1} + \eta_k,$$

where $\eta_k \sim N(0, V^2(\psi))$. With this setting, we can now estimate and predict the interest rates by means of the Kalman filter recursion. The main idea lies in writing down the log-likelihood function of all the R's via the Kalman equation in a recursive manner and optimize this likelihood function numerically. Before doing that, we present some basic data summaries in Table 11.1.

To get a feeling about the data we can import them into SPLUS as follows:

```
> zeros_read.table('zeros.dat',row.names=NULL,header=T)
```

Table 11.1 Summary Statistics

$r(t)$	Mean	Stand. Dev.
3-month	6.93597	1.0570630
6-month	7.102134	0.9762329
1-year	7.385955	0.9251570
2-year	7.723090	0.8190899
3-year	7.914313	0.7738133
5-year	8.146104	0.7417155
7-year	8.330224	0.7380349
10-year	8.484090	0.7152038

By using the following commands, we obtain a figure of the yield curve for February 91:

```
> maturity_c(0.25,0.5,1,2,3,5,7,10)
> plot(maturity,zeros$Feb91,type='b',xlab="maturity(years)",
+ ylab="rate",axes=F)
> axis(1,at=maturity,labels=as.character(maturity))
> axis(2)
> box(bty = "l")
```

Unfortunately, SPLUS cannot do much more for us in terms of using the Kalman filter, so we have to use some other package. There are available Kalman filter routines in MatLab as well as IMSL, which is a collection of numerical and statistical libraries written in Fortran, and it can be called from both Fortran and C. In particular, the KALMN routine performs the standard Kalman filter, and it can be used in conjunction with the UMINF routine to obtain maximum likelihood estimates using a quasi-Newton optimization method. It is stored in the file ims12.f on the Web page for this book.

The Fortran code follows:

```
c*****************************************************************
c MAIN PROGRAM, CALLS THE SUBROUTINES UMINF AND FUNC
c*****************************************************************
      program kalman
      integer nobs, nparam
      parameter (nobs=67,nparam=4)
c
      integer iparam(7)
      real func,fscale,fvalue,rparam(7),param(nparam),xguess
&          (nparam),xscale(nparam),ydata(8,67),r(8,8),covv(8,8)
      common ydata,r,covv
      external func,uminf,hfun,wfun,rfun
```

```
C*****************************************************************
c IMPORT THE DATA AS WELL AS THEIR SAMPLE VARIANCE COVARIANCE
c MATRIX
C*****************************************************************
      open(5,file='zeros.dat')
      read(5,*) ((ydata(i,j),j=1,67),i=1,8)

      open(50,file='varcovar')
      read(50,*) ((r(i,j),j=1,8),i=1,8)
C*****************************************************************
c STARTING VALUES FOR THE LIKELIHOOD MAXIMIZATION AS WELL AS
c PARAMETERS NEEDED IN UMINF
C*****************************************************************
      data xguess/7.0,70.0,15.0,2.0/,xscale/1.0,1.0,1.0,1.0/,
&          fscale/1.0/
c
      iparam(1)=0
      call uminf(func,nparam,xguess,xscale,fscale,iparam,
&          rparam,param,fvalue)
C*****************************************************************
c STANDARD OUTPUT
C*****************************************************************
      write(*,*) ' '
      write(*,*) '* * * Final estimates for Psi * * *'
      write(*,*) 'mu = ',param(1)
      write(*,*) 'theta = ',param(2)
      write(*,*) 'ksi = ',param(3)
      write(*,*) 'sigma = ',param(4)
      write(*,*) ' '
      write(*,*) '* * * optimization notes * * * '
      write(*,*) 'the number of iterations is ', iparam(3)
      write(*,*) 'and the number of function evaluations is ',
&          iparam(4)
      do 30 i = 1,8
         write(*,*) 'prediction standard errors  = ', sqrt(covv
&          (i,i))
 30   continue
      end


C*****************************************************************
c SUBROUTINE FUNC USED BY UMINF
c CALLS KALMN
C*****************************************************************
      subroutine func (nparam,param,ff)
      integer nparam
```

```
      real param(nparam), ff
c
      integer    ldcovb,ldcovv,ldq,ldr,ldt,ldz,nb,nobs,ny
      parameter (nb=2,nobs=67,ny=8,ldcovb=2,ldcovv=8,ldq=2,
     &       ldr=8,ldt=2,ldz=8)
c
      integer i,iq,it,n
      real alog,alndet,b(nb),covb(ldcovb,nb),covv(ldcovv,ny),
     &       q(ldq,nb),ss,t(ldt,nb),tol,v(ny),y(ny),ydata(ny,
     &       nobs),r(ldr,ny),z(ldz,nb),tau(ny),ksi(ny),phi
      common ydata,r,covv
      intrinsic alog
      external amach,kalmn,hfun,rfun,wfun
c
      data tau/3.0,6.0,12.0,24.0,36.0,60.0,84.0,120.0/
c
      tol=100.0*amach(4)

      do 5 i = 1,ny
         ksi(i) = param(3)*tau(i)
         z(i,1) = rfun(param,4) - wfun(param,4,tau(i))
         z(i,2) = -1.0*hfun(ksi(i))
  5   continue
c
      covb(1,1) = 0.0
      covb(1,2) = 0.0
      covb(2,1) = 0.0
      covb(2,2) = 1.0
c
      do 6 i =1,nb
         do 7 j=1,nb
            q(i,j) = covb(i,j)
  7      continue
  6   continue
c
      b(1) = 1.0
      b(2) = 0.0
      n=0
      ss=0
      alndet=0
      iq=0
      it=0
c
      phi = exp(-1.0*param(3)/nobs)
      t(1,1) = 1.0
```

```
      t(1,2) = 0.0
      t(2,1) = 0.0
      t(2,2) = phi

c

      do 10 i = 1,nobs
         y(1) = ydata(1,i)
         y(2) = ydata(2,i)
         y(3) = ydata(3,i)
         y(4) = ydata(4,i)
         y(5) = ydata(5,i)
         y(6) = ydata(6,i)
         y(7) = ydata(7,i)
         y(8) = ydata(8,i)
         call kalmn(ny,y,nb,z,ldz,r,ldr,it,t,ldt,iq,q,ldq,tol,
     &        b,covb,ldcovb,n,ss,alndet,v,covv,ldcovv)
  10     continue
      ff=n*alog(ss/n) + alndet
      return
      end
c*****************************************************************
c UTILITY FUNCTIONS
c*****************************************************************
      real function hfun(x)
      real x
c
      hfun = (1 - exp(-x))/x
      return
      end
c
c
      real function rfun(x,n)
      integer n
      real x(n)
c
      rfun = x(1) + x(2)*(x(4)/x(3)) -0.5*((x(4)**2)/(x(3)**2))
      return
      end
c
c
      real function wfun(x,n,tau)
      integer n
      real x(n),tau
c
      external hfun
```

```
       real ksit,ksit2
       ksit = x(3)*tau
       ksit2 = 2.0*ksit
       wfun = hfun(ksit)*(x(2)*(x(4)/x(3)) - ((x(4)**2)/(x(3)**2
&            ))) + 0.5*hfun(ksit2)*((x(4)**2)/(x(3)**2))
       return
       end
```

The output is given below:

```
* * * Final estimates for Psi * * *
mu =   5.05167
theta =   65.71723
ksi =   44.0575
sigma =   2.03501

* * * optimization notes * * *
the number of iterations is   11
and the number of function evaluations is   30
prediction standard errors  =  1.0571
prediction standard errors  =  .9762429
prediction standard errors  =  .9251597
prediction standard errors  =  .8190907
prediction standard errors  =  .7738137
prediction standard errors  =  .7417156
prediction standard errors  =  .7380349
prediction standard errors  =  .7152039
```

The estimated values of $\mu = 5.05167$, $\theta = 65.7172$, and $\sigma = 2.03501$ are very close to the results that Babbs and Nowman report ($\mu = 5.94$, $\theta = 64.83$, and $\sigma = 1.32$). Our estimated value for $\xi = 44.0575$, however, is somewhat different from the value reported there ($\xi = 19.1$). There could be many reasons for this discrepancy. First, the data sets are different since we do not use any swaption data in our example. Second, the time periods of the two different data sets are also different. Nevertheless, we see that the Kalman filter provides an efficient algorithm to estimate the parameters of a term-structure model.

11.6   EXERCISES

1. Consider the simple linear regression model
$$y_t = z_t b + w_t, \quad w_t \sim N(0.\sigma^2), \quad t = 1, 2, \dots .$$

   (a) Express this regression model in a state space form. Notice that the predictor z_t is known. Identify all the necessary coefficient matrices and errors in your state space form.

(b) Let $\hat{b}_n$ denote the least squares estimate of the regerssion coefficient b evaluated at time n. By means of direct computation, show that

$$\hat{b}_n = \hat{b}_{n-1} + K_n(y_n - z_n\hat{b}_{n-1})$$

for an appropriately defined Kalman gain matrix K_n and identify K_n.

(c) Compare your result with the Kalman filter recursions and identify Δ_n, G_n, and Ω_n.

(d) Show that
$$\text{var}\,(\hat{b}_n) = (1 - K_nz_n)\,\text{var}\,(\hat{b}_{n-1}).$$

Again, compare your formula with the Kalman filter recursions.

2. Consider the local trend model

$$Y_t = M_t + W_t, \quad W_t \sim \text{WN}(0, \sigma_w^2),$$
$$M_{t+1} = M_t + V_t, \quad V_t \sim \text{WN}(0, \sigma_v^2).$$

(a) Show that Y_t can be written as an ARIMA(0,1,1) model

$$Y_t = Y_{t-1} + Z_t + \theta Z_{t-1}$$

for an appropriately chosen noise sequence Z_t, where $Z_t \sim \text{WN}(0, \sigma^2)$.

(b) Determine θ and σ^2 in terms of σ_w^2 and σ_v^2.

(c) By letting $\sigma_w^2 = 8$ and $\sigma_v^2 = 20$, simulate a series of Y_t according to this local trend model.

(d) Perform a time series analysis on the simulated Y_t. What conclusions can you deduce?

3. Let Y_t be an ARMA(p, q) model with $p = 1$ and $q = 2$ defined by

$$(1 - \phi_1 B)Y_t = (1 - \theta_1 B - \theta_2 B^2)Z_t, \quad Z_t \sim \text{WN}(0, \sigma^2).$$

(a) By letting $r = \max\{p, q + 1\} = 3$, $\phi_j = 0$ for $j > 1$, $\theta_j = 0$ for $j > 2$, and $\theta_0 = 1$, show that X_t can be written in a state space form with the state vector defined by $\boldsymbol{X}_t = (X_{t-2}, X_{t-1}, X_t)'$, where X_t satisfies the AR(1) model $X_t = \phi_1 X_{t-1} + Z_t$. Consider a state space system with observation equation

$$Y_t = (-\theta_2, -\theta_1, \theta_0)' \boldsymbol{X}_t \tag{11.9}$$

and state equation

$$\boldsymbol{X}_{t+1} = \begin{pmatrix} 0 & 1 & 0 \\ 0 & 0 & 1 \\ \phi_3 & \phi_2 & \phi_1 \end{pmatrix} \boldsymbol{X}_t + \begin{pmatrix} 0 \\ 0 \\ 1 \end{pmatrix} Z_{t+1}.$$

Identify the system matrices $F, G, R,$ and Q. Show that the Y_t defined by equation (11.9) follows the ARMA(1,2) model given; that is, show that Y_t actually satisfies

$$(1 - \phi_1 B)Y_t = (1 - \theta_1 B - \theta_2 B^2)Z_t.$$

(b) Alternatively, let $m = \max\{p, q\} = 2$, $\phi_j = 0$ for $j > 1$, and let $\boldsymbol{X}_t$ be a two-dimensional state vector that satisfies the state equation

$$\boldsymbol{X}_{t+1} = \begin{pmatrix} 0 & 1 \\ \phi_2 & \phi_1 \end{pmatrix} \boldsymbol{X}_t + H Z_t,$$

where $H = (\psi_1, \psi_2)'$, ψ_1, and ψ_2 are the coefficients of z and z^2 in the expansion of $(1 - \theta_1 z - \theta_2 z^2)/(1 - \phi_1 z)$. Let

$$Y_t = (1, 0)\boldsymbol{X}_t + Z_t.$$

(c) Solve for H in terms of ϕ_1, θ_1, and θ_2 explicitly.

(d) Identify the system matrices F and G and deduce that $F^2 - \phi_1 F = \boldsymbol{0}$.

(e) By writing

$$\boldsymbol{X}_{t+1} = F\boldsymbol{X}_t + H Z_t,$$
$$Y_t = G\boldsymbol{X}_t + Z_t, \tag{11.10}$$

show that the Y_t defined in this way actually follows an ARMA(2,1) model; that is, deduce that the Y_t defined in equation (11.10) satisfies

$$(1 - \phi_1 B)Y_t = (1 - \theta_1 B - \theta_2 B^2)Z_t.$$

This is known as the *canonical observable representation* of an ARMA process.

(f) Which of the two state space formulations in parts (a) and (b) would you prefer? Comment briefly.

12

Multivariate GARCH

12.1   INTRODUCTION

Based on the discussions about multivariate ARMA models, it becomes very natural to consider modeling the volatility process in a higher-dimensional situation. So far, we have only considered the volatility of single asset returns and have tried to model the volatility process in terms of a univariate GARCH model or a univariate stochastic volatility model. Recall that for a univariate GARCH, the underlying equation is governed by

$$X_t = \sigma_t \epsilon_t, \quad \epsilon_t \sim \text{WN}(0,1),$$

where $\sigma_t^2 = E(X_t^2 | \mathcal{F}_{t-1})$ denotes the conditional variance that satisfies the equation

$$\sigma_t^2 = \alpha_o + \sum_{i=1}^{q} \alpha_i X_{t-i}^2 + \sum_{j=1}^{p} \beta_j \sigma_{t-j}^2.$$

In general, we may want to consider a portfolio that consists a vector of asset returns whose conditional covariance matrix evolves through time. Suppose that after filtering this multivariate series through ARMA models, we arrive at a portfolio that consists of k assets of return innovations $X_{i,t}$, $i = 1, \ldots, k$. Stacking these innovations into a vector $\boldsymbol{X}_t$, we define $\sigma_{ii,t} = \text{var}(X_{i,t} | \mathcal{F}_{t-1})$ and $\sigma_{ij,t} = \text{cov}(X_{i,t}, X_{j,t} | \mathcal{F}_{t-1})$. In this case, $\boldsymbol{\Sigma}_t = [\sigma_{ij,t}]$ denotes the conditional variance-covariance matrix of all the returns. The simplest generalization of the univariate GARCH(1,1) model relates the conditional variance-covariance matrix $\boldsymbol{\Sigma}_t$ to $\boldsymbol{X}_t\boldsymbol{X}_t'$ as follows:

$$\boldsymbol{X}_t = \boldsymbol{\Sigma}_t^{1/2}\boldsymbol{Z}_t,$$

$$E(\boldsymbol{X}_t\boldsymbol{X}_t'|\mathcal{F}_{t-1}) = \boldsymbol{\Sigma}_t.$$

One main difficulty encountered in this equation lies in finding a suitable system that describes the dynamics $\boldsymbol{\Sigma}_t$ parsimoniously. Without imposing further simplifications, it is easily seen that the model will become unmanageable rather quickly. Furthermore, the multiple GARCH equation also needs to satisfy the positive definiteness of $\boldsymbol{\Sigma}_t$. To illustrate the key ideas and to avoid unnecessarily cumbersome notation, unless otherwise specified, we shall restrict our attention to a multivariate GARCH(1,1) model with $k = 3$ for the remainder of this chapter.

12.2   GENERAL MODEL

To begin, we introduce the vech operator as follows. Given any square matrix A, vech A stacks elements on and below the main diagonal of A as follows:

$$\text{vech}\begin{pmatrix} a_{11} & a_{12} & a_{13} \\ a_{21} & a_{22} & a_{23} \\ a_{31} & a_{32} & a_{33} \end{pmatrix} = \begin{pmatrix} a_{11} \\ a_{21} \\ a_{31} \\ a_{22} \\ a_{32} \\ a_{33} \end{pmatrix}.$$

In general, if $\boldsymbol{A}$ is an $m \times m$ matrix, $\text{vech}(A)$ is an $m(m + 1)/2$-dimensional vector. The vech operator is usually applied to symmetric matrices in order to separate elements only. With this notation, we can model $\boldsymbol{\Sigma}_t$ as follows:

$$\text{vech}(\boldsymbol{\Sigma}_t) = \boldsymbol{\omega} + \boldsymbol{\Psi}\,\text{vech}(\boldsymbol{\Sigma}_{t-1}) + \boldsymbol{\Lambda}\,\text{vech}(\boldsymbol{X}_{t-1}\boldsymbol{X}_{t-1}'). \tag{12.1}$$

Note that $\boldsymbol{X}_t$, $\boldsymbol{Z}_t$, and $\boldsymbol{\omega}$ are all 3×1 vectors, $\boldsymbol{\omega}$ is a 6×1 vector, and $\boldsymbol{\Psi}$ and $\boldsymbol{\Lambda}$ are 6×6 square matrices. The total number of parameters in this model is

$$2\left[\frac{k(k + 1)}{2}\right]^2 + \frac{k(k + 1)}{2} = 108.$$

Even for moderate k, it is clear that this model can become very complicated. Certain restrictions need to be imposed on (12.1) to reduce the number of parameters and to ensure that $\boldsymbol{\Sigma}_t$ is positive definite.

Example 12.1 *To get an idea as to how the dynamics of volatilities are described by* (12.1), *consider the special case where* $\boldsymbol{\omega} = 0$ *and* $\boldsymbol{\Psi} = \mathbf{0}$:

$$
\text{vech}(\boldsymbol{\Sigma}_t) = \begin{pmatrix} \sigma_{11,t} \\ \sigma_{21,t} \\ \sigma_{22,t} \\ \sigma_{31,t} \\ \sigma_{32,t} \\ \sigma_{33,t} \end{pmatrix} = \begin{pmatrix} \lambda_{11} & \lambda_{12} & \cdots & \lambda_{15} & \lambda_{16} \\ \lambda_{21} & \lambda_{22} & \cdots & \lambda_{25} & \lambda_{26} \\ \lambda_{31} & \lambda_{32} & \cdots & \lambda_{35} & \lambda_{36} \\ \lambda_{41} & \lambda_{42} & \cdots & \lambda_{45} & \lambda_{46} \\ \lambda_{51} & \lambda_{52} & \cdots & \lambda_{55} & \lambda_{56} \\ \lambda_{61} & \lambda_{62} & \cdots & \lambda_{65} & \lambda_{66} \end{pmatrix} \begin{pmatrix} X_{1,t-1}X_{1,t-1} \\ X_{2,t-1}X_{1,t-1} \\ X_{2,t-1}X_{2,t-1} \\ X_{3,t-1}X_{1,t-1} \\ X_{3,t-1}X_{2,t-1} \\ X_{3,t-1}X_{3,t-1} \end{pmatrix}.
$$

When equating the elements on the two sides of the preceding equation, we see that each volatility is related to past squares of returns in a rather complicated manner. For example, writing out the first equation gives

$$
\sigma_{11,t} = \lambda_{11}X_{1,t-1}^2 + \lambda_{12}X_{2,t-1}X_{1,t-1} + \lambda_{13}X_{2,t-1}^2
$$
$$
+ \lambda_{14}X_{3,t-1}X_{1,t-1} + \lambda_{15}X_{3,t-1}X_{2,t-1} + \lambda_{16}X_{3,t-1}^2.
$$

Even under the simplifying assumptions that $\boldsymbol{\Sigma}_t$ *depends only on* $\boldsymbol{X}_t$, *but not* $\boldsymbol{\omega}$ *or* $\boldsymbol{\Sigma}_{t-1}$ *(i.e.,* $\boldsymbol{\omega} = 0$ *and* $\boldsymbol{\Psi} = \mathbf{0}$*), this equation is far from being simple.* □

12.2.1   Diagonal Form

The first simple model of (12.1) is the case when $\boldsymbol{\Psi}$ and $\boldsymbol{\Lambda}$ are both diagonal matrices. In this case the (i,j)th element of $\boldsymbol{\Sigma}_t$ is

$$
\sigma_{ij,t} = \omega_{ij} + \beta_{ij}\sigma_{ij,t-1} + \alpha_{ij}X_{i,t-1}X_{j,t-1}. \tag{12.2}
$$

For this model, each element of $\boldsymbol{\Sigma}_t$ follows a univariate GARCH(1,1) model driven by the corresponding elements of the cross-product matrix $\boldsymbol{X}_{t-1}\boldsymbol{X}_{t-1}'$ and the element $\sigma_{ij,t-1}$. This model has three parameters for each element of $\boldsymbol{\Sigma}_t$ and thus has $3k(k+1)/2 = 18$ parameters for the entire model.

Example 12.2 *Continuing with Example 12.1, where* $\boldsymbol{\omega} = 0$ *and* $\boldsymbol{\Psi} = \mathbf{0}$, *the matrix* $\boldsymbol{\Lambda} = \text{diag}(\lambda_1, \ldots, \lambda_6)$. *In this case, the elements* α_{ij} *in* (12.2) *are related to the diagonal elements through the equation* $(\lambda_1, \lambda_2, \lambda_3, \lambda_4, \lambda_5, \lambda_6) = (\alpha_{11}, \alpha_{21}, \alpha_{22}, \alpha_{31}, \alpha_{32}, \alpha_{33})$.

In SPLUS, *this model is known as the vector-diagonal model. Specifically, let a three-dimensional vector* $\boldsymbol{a} = (a_1, a_2, a_3)'$. *Then an equivalent vectorize version becomes*

$$
\text{vech}(\boldsymbol{\Sigma}_t) = \text{diag}(\text{vech}(\boldsymbol{aa}^T))\,\text{vech}(\boldsymbol{X}_{t-1}\boldsymbol{X}_{t-1}^T).
$$

Note that this is a special case of the diagonal form with $\boldsymbol{\Lambda} = \text{diag}(\text{vech}(\boldsymbol{aa}^T))$. □

12.2.2   Alternative Matrix Form

Instead of vectorizing Σ_t, we may want to model it directly in a matrix form. There appears to be no unique way of doing this. In SPLUS it uses the form

$$\Sigma_t = \omega + \sum_{i=1}^{q} A_i \otimes (X_{t-i} X_{t-i}^T) + \sum_{i=1}^{p} B_i \otimes \Sigma_{t-i}, \qquad (12.3)$$

where the symbol $\otimes$ stands for the Hadamard product (element-by-element multiplication). All quantities in equation (12.3) are $k \times k$ matrices (k is the dimension of the observed vector) except for X_t, which is a $k \times 1$ column vector. The matrices ω, A_i, and B_i must be *symmetric*.

For the GARCH(1,1) case, equation (12.3) reduces to

$$\Sigma_t = \omega + A \otimes (X_{t-1} X_{t-1}^T) + B \otimes \Sigma_{t-1}.$$

Consider again the simple example where $\omega = B = 0$. In the diagonal form (12.2), this equation becomes

$$\Sigma_t = A \otimes (X_{t-1} X_{t-1}^T),$$

where A is symmetric and $\text{vech}(A) = \Lambda = \text{diag}(\lambda_1, \dots, \lambda_6)$. In the particular case that $A = (aa^T)$, where $a = (a_1, a_2, a_3)'$, the diagonal form studied in (12.2) becomes

$$\Sigma_t = (aa^T) \otimes (X_{t-1} X_{t-1}^T).$$

12.3   QUADRATIC FORM

A model termed the BEKK model by Engle and Kroner (1995) works with quadratic forms rather than individual elements of Σ_t. In this case we write

$$\Sigma_t = C'C + B'\Sigma_{t-1}B + A'X_{t-1}X_{t-1}'A, \qquad (12.4)$$

where C is a lower triangular matrix with $k(k+1)/2$ parameters, and B and A are $k \times k$ square matrices with k^2 parameters, giving rise to a total of $2k^2 + k(k+1)/2 = (5k^2 + k)/2 = 24$ parameters. Weak restrictions on B and A guarantee that Σ_t is always positive definite. Again, SPLUS allows for this specification, known as the *BEKK model*. In this case, the matrices A and B do not necessarily have to be symmetric.

12.3.1   Single-Factor GARCH(1,1)

A special case of the quadratic form or BEKK model is known as the *single-factor model*. In this case we try to model the volatilities among the assets in terms of one single source (a factor). Specifically, we write

$$\Sigma_t = C'C + \lambda\lambda'(\beta w'\Sigma_{t-1}w + \alpha(w'X_{t-1})^2), \qquad (12.5)$$

where λ and w are k-dimensional vectors and α and β are scalars. C is again a lower triangular matrix with $k(k+1)/2$ parameters. Usually, we assume that $w = (w_1, \ldots, w_k)'$ such that $\sum_i w_i = 1$ (i.e., we think of w as weights). Let $X_{pt} = w'X_t$ and $\sigma_{pp,t} = w'\Sigma_t w$. Then the (i,j)th element of Σ_t can be written as

$$\sigma_{ij,t} = c_{ij} + \lambda_i \lambda_j \beta \sigma_{pp,t-1} + \lambda_i \lambda_j \alpha X^2_{p,t-1},$$

where c_{ij} is the (i,j)th element of the matrix $C'C$. Consider the expression

$$\sigma_{pp,t} = w'\Sigma_{t-1}w = w'C'Cw + w'\lambda\lambda'w(\beta w'\Sigma_{t-1}w + \alpha(w'X_{t-1})^2)$$

$$= (Cw)'(Cw) + (\lambda'w)^2(\beta\sigma_{pp,t-1} + \alpha X^2_{p,t-1}).$$

We deduce that

$$\sigma_{ij,t} = w_{ij} + \tilde{\lambda}_i \tilde{\lambda}_j \sigma_{pp,t},$$

$$\sigma_{pp,t} = w_{pp} + \tilde{\beta}\sigma_{pp,t-1} + \tilde{\alpha} X^2_{p,t-1},$$

where $\tilde{\lambda}_i = \lambda_i/(\lambda'w)^2$, $\tilde{\alpha} = (\lambda'w)^2\alpha$, $\tilde{\beta} = (\lambda'w)^2\beta$, with w_{ij}, and w_{pp} defined appropriately. In this setting, the covariance of any two assets (returns) move through time only with the variance of the portfolio, $\sigma_{pp,t}$, and this variance follows a univariate GARCH(1,1). It can easily be seen that this is a special case of the BEKK model where the matrices in the quadratic form $A = \sqrt{\alpha}\, w\lambda'$ and $B = \sqrt{\beta}\, w\lambda'$. It has $(k^2 + 5k + 2)/2 = 13$ parameters. Since this is a special case of the BEKK model, SPLUS can be used to fit this model.

12.3.2   Constant-Correlation Model

In this model, each return variance follows a GARCH(1,1) model and the covariance between any two returns is given by a constant correlation times the conditional standard deviation. Specifically, we write

$$\sigma_{ii,t} = w_{ii} + \beta_{ii}\sigma_{ii,t-1} + \alpha_{ii}X^2_{t-1,i},$$

$$\sigma_{ij,t} = \rho_{ij}(\sigma_{ii,t}\sigma_{jj,t})^{1/2}.$$

This model has a total of k α's, k β's, and $k(k+1)/2$ ρ's giving a total of $k(k+5)/2 = 12$ parameters. In SPLUS this model is given by the command `ccc.g(p,q)`. We illustrate these arrays of models with SPLUS in the next section.

12.4   EXAMPLE OF FOREIGN EXCHANGE RATES

In this example we study the foreign exchange rates used in Diebold and Nerlove (1989). Although Diebold and Nerlove use a single-factor ARCH model, we do not follow their approach completely. Instead, we use the data as a vehicle to illustrate the various features of SPLUS in fitting a multivariate GARCH model.

12.4.1   The Data

We study seven major dollar spot rates; the Canadian dollar, deutsche mark, British pound, Japanese yen, French franc, Italian lira, and Swiss franc. The data are from the first week of July 1973 to the second week of August 1985 and represent seasonally unadjusted interbank closing spot prices on Wednesdays, taken from the *International Monetary Markets Yearbook*. They are available on the Web page for this book under the file `forex.dat`. We read the data in SPLUS using the `read.table` command, and convert them into regular time series as follows:

```
> forex_read.table('forex.dat',row.names=NULL,header=T)
> forex2_rts(forex[,2:8],start=c(1973,29),freq=52)
```

12.4.2   Multivariate GARCH in SPLUS

Following the foregoing discussion, we restrict our attention to three components, $k = 3$. We deal with the deutsche mark (DM), the British pound (BP), and the Japanese yen (YEN) only, denoted by $S_t = (S_{1t}, S_{2t}, S_{3t})'$. Their time series plots are shown in Figure 12.1.

A visual inspection indicates nonstationarity in each of the series, so we proceed by taking differenced natural logs [i.e., $\boldsymbol{X}_t = (\boldsymbol{I} - \boldsymbol{B}) \log S_t$], which allow convenient interpretation by means of the approximate percentage change. The time series plots of $\boldsymbol{X}_t$ are given in Figure 12.2.

```
> dlmpy_diff(log(mpy))
> tsplot(dlmpy)
> legend(100,0.06,legend=names(as.data.frame(dlmpy)),lty=1:3)
```

After taking this nonstationarity into account, we can use the function `mgarch` to fit multivariate GARCH models. One of the most general models handled is the *diagonal-vec form* discussed in Section 3.2.2. In our case where $p = q = 1$, the conditional variance matrix is given by

$$\boldsymbol{\Sigma}_t = \boldsymbol{\omega} + \boldsymbol{A}_1 \otimes (\boldsymbol{X}_{t-1}\boldsymbol{X}_{t-1}^T) + \boldsymbol{B}_1 \otimes \boldsymbol{\Sigma}_{t-1}, \tag{12.6}$$

where the symbol $\otimes$ stands for the Hadamard product (element-by-element multiplication). All quantities in equation (12.6) are 3×3 matrices except for $\boldsymbol{X}_t$, which is a 3×1 column vector. The matrices $\boldsymbol{\omega}$, $\boldsymbol{A}_1$, and $\boldsymbol{B}_1$ must be *symmetric*. Notice that this form models the conditional covariance matrix in an element-by-element manner. The command

```
series.mod_mgarch(series~-1, ~dvec(1,1))
```

will fit the model shown in equation (12.6) for $p = 1$ and $q = 1$. This model is also referred to as the *order* (1,1) *diagonal-vec model*.

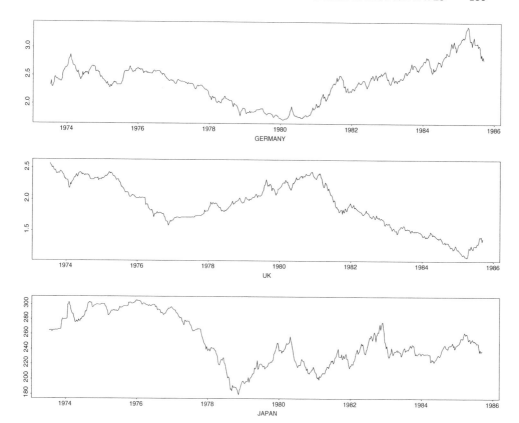

Fig. 12.1 Time series plots of the raw data.

Example 12.3 *We use the function* mgarch *to fit the order* $(1,1)$ *diagonal-vec model*:

```
> mpy.mod.dvec_mgarch(dlmpy ~-1, ~dvec(1,1))
```

To see a brief display of the model and the estimated coefficient values, type

```
> mpy.mod.dvec

Call: mgarch(formula.mean = dlmpy~-1, formula.var=~dvec(1,1))

Mean Equation: dlmpy ~ -1

Conditional Variance Equation:   ~ dvec(1,1)
```

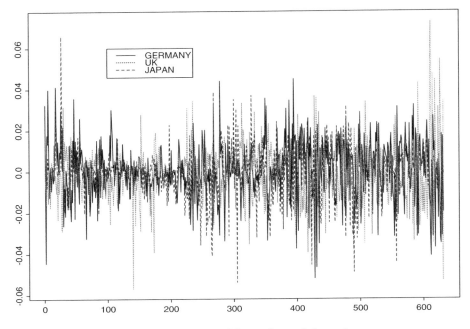

Fig. 12.2   Differenced logarithms of the series.

Coefficients:

```
        A(1, 1)   1.199e-05
        A(2, 1)  -4.955e-06
        A(3, 1)   1.066e-05
        A(2, 2)   2.207e-06
        A(3, 2)  -1.724e-06
        A(3, 3)   1.235e-06
 ARCH(1; 1, 1)    1.070e-01
 ARCH(1; 2, 1)    1.090e-01
 ARCH(1; 3, 1)    1.032e-01
 ARCH(1; 2, 2)    6.910e-02
 ARCH(1; 3, 2)    7.747e-02
 ARCH(1; 3, 3)    5.348e-02
GARCH(1; 1, 1)    8.211e-01
GARCH(1; 2, 1)    8.344e-01
GARCH(1; 3, 1)    8.385e-01
GARCH(1; 2, 2)    9.000e-01
GARCH(1; 3, 2)    8.748e-01
GARCH(1; 3, 3)    9.393e-01
```

From the output above, the estimated coefficients in equation (12.6) are

$$\omega = \begin{pmatrix} 1.199e\text{-}05 & -4.955e\text{-}06 & 1.066e\text{-}05 \\ -4.955e\text{-}06 & 2.207e\text{-}06 & -1.724e\text{-}06 \\ 1.066e\text{-}05 & -1.724e\text{-}06 & 1.235e\text{-}06 \end{pmatrix},$$

$$\boldsymbol{B}_1 = \begin{pmatrix} 8.211e\text{-}01 & 8.344e\text{-}01 & 8.385e\text{-}01 \\ 8.344e\text{-}01 & 9.000e\text{-}01 & 8.748e\text{-}01 \\ 8.385e\text{-}01 & 8.748e\text{-}01 & 9.393e\text{-}01 \end{pmatrix},$$

and

$$\boldsymbol{A}_1 = \begin{pmatrix} 1.070e\text{-}01 & 1.090e\text{-}01 & 1.032e\text{-}01 \\ 1.090e\text{-}01 & 6.910e\text{-}02 & 7.747e\text{-}02 \\ 1.032e\text{-}01 & 7.747e\text{-}02 & 5.348e\text{-}02 \end{pmatrix}.$$

The function summary *provides not only the information above, but also standard inference results about the parameter estimates and test statistics :*

```
> summary(mpy.mod.dvec)

Call: mgarch(formula.mean=dlmpy~-1, formula.var=~dvec(1,1))

Mean Equation: dlmpy ~ -1

Conditional Variance Equation:   ~ dvec(1,1)

Conditional Distribution:  gaussian


-------------------------------------------------------
Estimated Coefficients:
-------------------------------------------------------
                   Value Std.Error  t value  Pr(>|t|)
       A(1, 1)  1.199e-05 2.660e-06    4.507 3.915e-06
       A(2, 1) -4.955e-06 1.438e-06   -3.446 3.036e-04
       A(3, 1)  1.066e-05 2.155e-06    4.946 4.878e-07
       A(2, 2)  2.207e-06 7.930e-07    2.784 2.769e-03
       A(3, 2) -1.724e-06 7.750e-07   -2.224 1.324e-02
       A(3, 3)  1.235e-06 4.002e-07    3.086 1.060e-03
  ARCH(1; 1, 1)  1.070e-01 1.463e-02    7.312 4.030e-13
  ARCH(1; 2, 1)  1.090e-01 1.454e-02    7.496 1.126e-13
  ARCH(1; 3, 1)  1.032e-01 1.908e-02    5.407 4.564e-08
  ARCH(1; 2, 2)  6.910e-02 1.161e-02    5.949 2.239e-09
  ARCH(1; 3, 2)  7.747e-02 1.304e-02    5.941 2.351e-09
  ARCH(1; 3, 3)  5.348e-02 6.215e-03    8.604 0.000e+00
```

```
GARCH(1; 1, 1)  8.211e-01 2.308e-02   35.569 0.000e+00
GARCH(1; 2, 1)  8.344e-01 1.983e-02   42.084 0.000e+00
GARCH(1; 3, 1)  8.385e-01 2.614e-02   32.074 0.000e+00
GARCH(1; 2, 2)  9.000e-01 1.599e-02   56.271 0.000e+00
GARCH(1; 3, 2)  8.748e-01 2.274e-02   38.472 0.000e+00
GARCH(1; 3, 3)  9.393e-01 5.588e-03  168.100 0.000e+00
```

```
AIC(18) = -11804.52
BIC(18) = -11724.44
```

Normality Test:

	Jarque-Bera	P-value	Shapiro-Wilk	P-value
GERMANY	14.76	0.0006226	0.9869	0.6746
UK	1203.23	0.0000000	0.9521	0.0000
JAPAN	876.58	0.0000000	0.9560	0.0000

Ljung-Box test for standardized residuals:

	Statistic	P-value	Chi-square d.f.
GERMANY	25.34	0.01331	12
UK	17.27	0.13965	12
JAPAN	17.71	0.12482	12

Ljung-Box test for squared standardized residuals:

	Statistic	P-value	Chi-square d.f.
GERMANY	8.433	0.7505	12
UK	1.433	0.9999	12
JAPAN	4.835	0.9633	12

Lagrange multiplier test:

	Lag 1	Lag 2	Lag 3	Lag 4	Lag 5
GERMANY	1.15796	1.224393	-0.48768	-0.5168	-0.6364
UK	-0.24630	-0.031264	0.06862	-0.3687	-0.3745
JAPAN	0.08048	-0.006353	-0.08270	-0.1131	-0.4491

	Lag 6	Lag 7	Lag 8
GERMANY	1.5077	-0.98599	-0.205
UK	-0.5422	-0.08647	-0.102
JAPAN	-0.5800	0.30541	1.154

	Lag 9	Lag 10	Lag 11	Lag 12	C
GERMANY	0.4657	0.4067	1.1327	-0.1457	0.66144
UK	0.1230	-0.1714	0.8290	-0.0212	0.09538
JAPAN	-0.4607	-0.3207	-0.8834	0.1006	1.15361

	TR^2	P-value	F-stat	P-value
GERMANY	9.112	0.6933	0.8407	0.7094
UK	1.440	0.9999	0.1312	1.0000
JAPAN	4.556	0.9712	0.4173	0.9935

To make visual assessment of the goodness of fit of the model mpy.mod.dvec, *we can use the* plot *function*

```
> plot(mpy.mod.dvec)

Make a plot selection (or 0 to exit):

1: plot: All
2: plot: Observations and ACFs
3: plot: ACFs of Cross-product of Observations
4: plot: ACF of the Squared Norm of the Observations
5: plot: Residuals and Conditional SDs
6: plot: Standardized Residuals and ACFs
7: plot: ACFs of Cross-product of Std Residuals
8: plot: ACF of the Squared Norm of the Std Residuals
9: plot: QQ-plots of the Standardized Residuals
```

□

Visual comparisons between the behavior of the multivariate time series and the behavior of the multivariate *standardized residuals* for the fitted model provide excellent guidance in assessing the fit of the model. Note here that while in the univariate case the standardized residuals are simply the model residuals divided by the conditional standard deviation, the standardized residuals for a multivariate GARCH are obtained by the *whitening matrix transformation*:

$$\Sigma_t^{-1/2} Z_t.$$

For a well-fitted multivariate GARCH model, this transformation produces standardized residual vectors that have an approximately diagonal conditional covariance matrix.

We can make a comparison plot as follows. Select menu choices 2 and 6 to obtain the plots in Figures 12.3 and 12.4. The fitted model has resulted in somewhat smaller ACF values for the standardized residuals relative to the series observed.

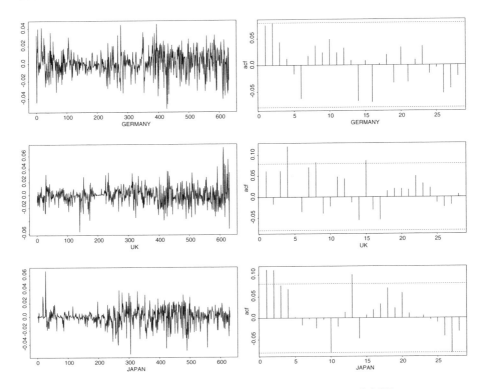

Fig. 12.3 Selecting menu choice 2: observations and ACFs.

Make menu selections 3 and 7 to get the ACFs of the cross-product plots shown in Figures 12.5 and 12.6. By the *ACFs of cross-products*, we mean the ACFs of all pairwise products of two or more multivariate time series. For our tri-variate series we get the following: The upper-left-hand plot is the ACF of the squared values of the DM series; the lower-right-hand plot is the ACF of the squared values of the YEN series and the lower-left-hand plot is the ACF of the product of the DM and YEN series. Figure 12.5 for the series observed reveals significant autocorrelation in all six plots, indicating a need for GARCH modeling.

Figure 12.6 for the standardized residuals series shows that most but not all of the autocorrelation structure has been removed by the fitted diagonal-vec GARCH model. We may wish to try and refine the trivariate GARCH model in order to remove the small number of remaining significant correlations in the ACFs of the standardized residuals.

When we have more than two series to model, it may be inconvenient to plot and compare the ACFs of all cross-products. SPLUS addresses this issue by providing plots for the ACFs of the Euclidean norm of the residuals and the standardized residuals, where the *Euclidean norm* is the square root of

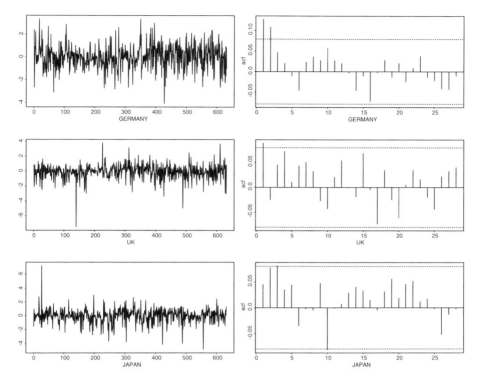

Fig. 12.4 Selecting menu choice 6: standardized residuals and ACFs.

the sum of squares of the elements of a vector. If the ACF of the norm of the multivariate series exhibits significant correlation, multivariate GARCH modeling may be called for. If the ACF of the norm of the standardized residuals for the fitted model exhibits little correlation, the GARCH model fit is probably pretty good. Menu selections 4 and 8 result in these ACF plots. We can also get paired plots of the residuals and their conditional standard deviations by making menu selection 5. Finally, by making menu selection 9, we can look at the normal Q–Q plots of the standardized residuals, which are shown in Figure 12.7.

The object `mpy.mod.dvec` has several components which can be extracted. Using `residuals` or `sigma.t` will return matrices with each column corresponding to one of the time series in the multivariate set. For example,

```
residuals(mpy.mod.dvec)
```

is a 632×3 matrix with the first column corresponding to the DM series of length 632, the second column corresponding to the BP series, and the third to the YEN series. We can extract the standardized residuals using the `standardize=T` option in the call of the function `residuals`.

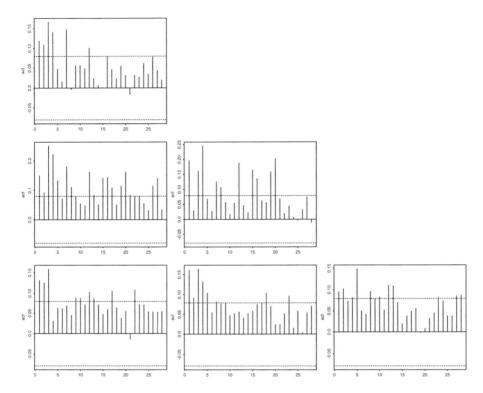

Fig. 12.5   Selecting menu choice 3: ACFs of cross-product observations.

We get Figures 12.8 and 12.9 using the commands

```
> pairs(dlmpy)
> pairs(residuals(mpy.mod.dvec,standardize=T))
```

As we can see in Figure 12.8, the data sets for the original series are highly correlated, with a few outliers in evidence. In Figure 12.9 we see that the bulk of the standardized bivariate residuals have a nearly circular scatter plot, which means that the standardization has resulted in the bulk of the residuals being uncorrelated. The few outliers seen in Figure 12.8 remain in Figure 12.9 but with different orientations.

12.4.3   Prediction

Once we fit a multivariate GARCH model, we can use the `predict` function, with the fitted model object as an argument, to create an object of the

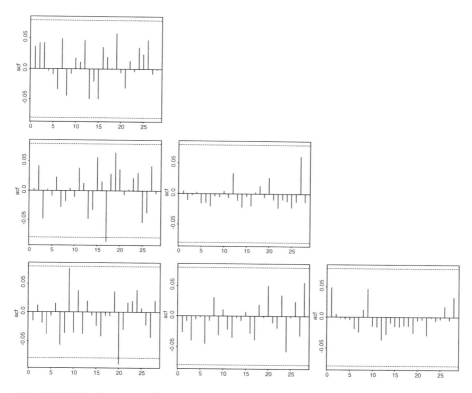

Fig. 12.6   Selecting menu choice 7: ACFs of cross-product of standardized residuals.

class *mgarch.predict*. This *predictions* object has the following components: `series.pred`, the predicted values of the series; `sigma.pred`, the predicted values of the conditional standard deviations; and `R.pred`, the predicted values of the conditional correlation matrix. After creating the *mgarch.predict* object, we can use the `plot` function to get plots of these predicted series.

12.4.4   Predicting Portfolio Conditional Standard Deviations

If we have a portfolio with weights $w = (w_1, \ldots, w_d)$, where d represents many financial time series returns, with the weights summing to 1, we can predict the conditional standard deviations of the portfolio y using arguments optional for the `plot` function. In our example, if the weights are $w = (0.25, 0.50, 0.25)$, we can use the following:

```
> mpy.mod.dvec.pred_predict(mpy.mod.dvec,20)
> plot(mpy.mod.dvec.pred,portfolio=T,weights=c(0.25,0.5,0.25))
```

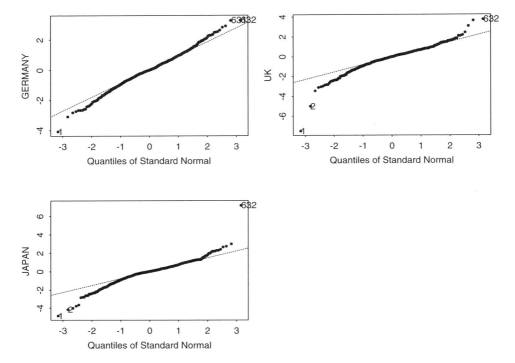

Fig. 12.7 Selecting menu choice 9: Q–Q plots of the standardized residuals.

12.4.5 BEKK Model

The quadratic model, also known as the BEKK model, can be fitted in SPLUS using `~bekk(p,q)` as the second argument to `mgarch`. For example:

```
> mpy.mod.bekk_mgarch(dlmpy ~-1, ~bekk(1,1),trace=F)
```

The BEKK(1,1) model has the following conditional covariance matrix structure:

$$\boldsymbol{\Sigma}_t = \boldsymbol{A}\boldsymbol{A}^T + \boldsymbol{A}_1(\boldsymbol{X}_{t-1}\boldsymbol{X}_{t-1}^T)\boldsymbol{A}_1^T + \boldsymbol{B}_1\boldsymbol{\Sigma}_{t-1}\boldsymbol{B}_1^T. \tag{12.7}$$

Because of the presence of a paired transposed matrix factor for each of the $d \times d$ matrices $\boldsymbol{A}$, $\boldsymbol{A}_1$, and $\boldsymbol{B}_1$, symmetry and nonnegative-definiteness of the conditional covariance matrix $\boldsymbol{\Sigma}_t$ are assured.

For our example, we have

$$\boldsymbol{A} = \begin{pmatrix} 1.890e{-}03 & 1.105e{-}04 & 3.330e{-}03 \\ 1.105e{-}04 & -1.469e{-}03 & 1.296e{-}04 \\ 3.330e{-}03 & 1.296e{-}04 & 6.779e{-}07 \end{pmatrix},$$

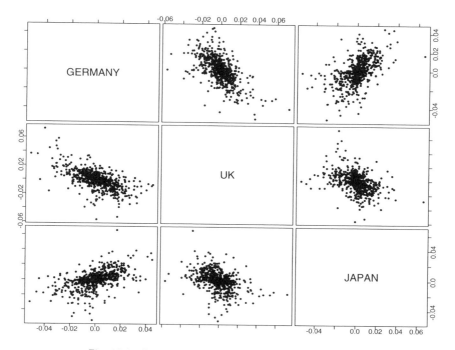

Fig. 12.8   Bivariate scatter plot of the original series.

$$A_1 = \begin{pmatrix} 9.187e\text{-}01 & -3.632e\text{-}02 & 6.526e\text{-}02 \\ -9.526e\text{-}04 & 8.925e\text{-}01 & 3.776e\text{-}02 \\ 4.472e\text{-}02 & -1.398e\text{-}02 & 9.339e\text{-}01 \end{pmatrix},$$

and

$$B_1 = \begin{pmatrix} 3.510e\text{-}01 & -1.005e\text{-}04 & -9.379e\text{-}02 \\ -8.426e\text{-}03 & 3.195e\text{-}01 & -5.906e\text{-}02 \\ -1.034e\text{-}01 & 1.873e\text{-}02 & 3.035e\text{-}01 \end{pmatrix}.$$

12.4.6   Vector-Diagonal Models

In *vector-diagonal* models we assume that $\mathbf{\Psi}$ and $\mathbf{\Lambda}$ in equation (12.6) are diagonal matrices. In this case we have

$$\mathbf{\Sigma}_t = \boldsymbol{A}\boldsymbol{A}^T + \boldsymbol{a}_1\boldsymbol{a}_1^T \otimes (\boldsymbol{X}_{t-1}\boldsymbol{X}_{t-1}^T) + \boldsymbol{b}_1\boldsymbol{b}_1^T \otimes \mathbf{\Sigma}_{t-1}. \qquad (12.8)$$

For this model, each element of $\mathbf{\Sigma}_t$ follows a univariate GARCH model driven by the corresponding element of the cross-product matrix $\boldsymbol{X}_{t-1}\boldsymbol{X}'_{t-1}$. This model is obtained by making the matrices $\mathbf{\Lambda}$ and $\mathbf{\Psi}$ diagonal.

The diagonal form, seen in (12.8), can also be fitted in SPLUS using the `~dvec.vec.vec(p,q)` argument. For example

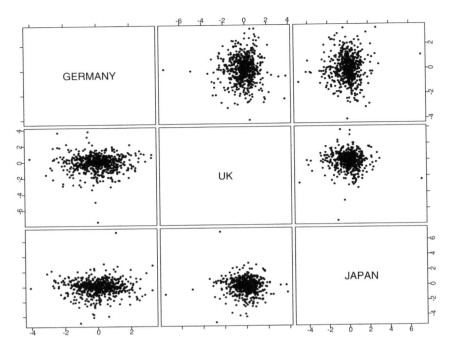

Fig. 12.9 Bivariate scatter plot of the standardized residuals.

```
> mpy.mod.vec_mgarch(dlmpy ~-1, ~dvec.vec.vec(1,1),trace=F)
```

leads us to the estimates

$$A = \begin{pmatrix} 2.606e\text{-}03 & -1.673e\text{-}03 & 2.860e\text{-}03 \\ -1.673e\text{-}03 & 7.679e\text{-}04 & -5.629e\text{-}05 \\ 2.860e\text{-}03 & -5.629e\text{-}05 & 8.097e\text{-}04 \end{pmatrix},$$

$$a_1 = \begin{pmatrix} 9.320e\text{-}01 \\ 9.152e\text{-}01 \\ 9.698e\text{-}01 \end{pmatrix},$$

and

$$b_1 = \begin{pmatrix} 2.991e\text{-}01 \\ 3.195e\text{-}01 \\ 2.292e\text{-}01 \end{pmatrix}.$$

12.4.7 ARMA in Conditional Mean

If there is a mean structure, we fit the following general model with a vector ARMA structure:

$$y_t = c + \sum_{i=1}^{r} \Phi_i y_{t-i} + \sum_{i=1}^{s} \Theta_i X_{t-1} + X_t. \tag{12.9}$$

For example, we fit a diagonal ARMA(1,1) with a dvec(1,1) model for the residuals by using the command

```
> mpy.mod.arma_mgarch(dlmpy ~ arma(1,1), ~dvec(1,1), trace=F)
```

In our example we get

$$c = \begin{pmatrix} -7.867e\text{-}05 \\ -3.914e\text{-}04 \\ -1.489e\text{-}04 \end{pmatrix},$$

$$\boldsymbol{\Phi}_1 = \text{diag}\{4.452e\text{-}01, 5.824e\text{-}01, 6.590e\text{-}01\},$$

$$\boldsymbol{\Theta}_1 = \text{diag}\{-3.532e\text{-}01, -4.830e\text{-}01, -5.998e\text{-}01\},$$

$$\boldsymbol{A}_1 = \begin{pmatrix} 8.365e\text{-}01 & 8.335e\text{-}01 & 8.309e\text{-}01 \\ 8.335e\text{-}01 & 9.141e\text{-}01 & 8.791e\text{-}01 \\ 8.309e\text{-}01 & 8.791e\text{-}01 & 9.402e\text{-}01 \end{pmatrix},$$

and

$$\boldsymbol{B}_1 = \begin{pmatrix} 9.904e\text{-}02 & 1.088e\text{-}01 & 1.072e\text{-}01 \\ 1.088e\text{-}01 & 6.085e\text{-}02 & 7.395e\text{-}02 \\ 1.072e\text{-}01 & 7.395e\text{-}02 & 5.342e\text{-}02 \end{pmatrix}.$$

12.5 CONCLUSIONS

In this chapter we have considered an array of multivariate GARCH models. Due to the high-dimensional nature of the problem, it becomes inevitable that many of the models used in fitting multivariate GARCH series possess a rather complicated form. It is hoped that by following the examples given in this chapter, readers can acquire the essentials of fitting some commonly used multivariate GARCH models. It should also be pointed out that at the time of writing of this book, there are very few software products that support multivariate GARCH models. The only two commonly used platforms are SPLUS and SAS. It is a rule rather than an exception that users must develop their own software support for fitting multivariate GARCH series.

12.6 EXERCISES

1. Consider the foreign exchange rate example analyzed in Diebold and Nerlove (1989) and Section 12.4.1 of this chapter. Download this data set from the Web page for this book.

   (a) Perform a univariate GARCH analysis on all seven countries as done in the paper. You might also want to fit an AR structure on the conditional mean as well. Comment on your analysis with respect to the findings given in Diebold and Nerlove.

(b) Perform a multivariate GARCH analysis on the following four countries: U.K., Germany, Canada, and Japan. You might want to fit a number of models, such as the BEKK or the diagonal model. Comment on your analysis.

(c) In particular, utilize a constant correlation model for these four countries. Comment on your results with the findings of the single-factor model given in Diebold and Nerlove.

2. Finding a reasonable theoretical assignment for this chapter is almost impossible. For cultural enrichment and entertainment, with the obvious modifications, one might want to consider the following celebrated assignment from Lang (1965). Take any article or book on GARCH models and prove all the theorems without looking at the proofs given in that article or book!

13

Cointegrations and Common Trends

13.1 INTRODUCTION

Cointegration deals with the common behavior of a multivariate time series. It often happens in practice that each component of a multivariate time series may be nonstationary, but certain linear combinations of these components are stationary. Cointegration studies the effects of these combinations and the relationships among the components.

A large number of texts have been written on cointegration since the publication of the seminal paper by Engle and Granger (1987). For example, the book by Johansen (1995) provides one of the most complete coverages of the statistical theory of cointegrated time series. A succinct survey of some of the results presented in this book is given in the review article by Johansen (1996). Other references about statistical inference for cointegrated systems include Stock and Watson (1988), Phillips (1991), Park (1992), Reinsel and Ahn (1992), and Chan and Tsay (1996).

On the econometric front, several monographs and special collections of papers have also been devoted to this subject; see, for example, Banerjee, Dolado, Galbraith, and Hendry (1993), Engle and Granger (1991), Maddala and Kim (1998), and special issues of *Oxford Bulletin of Economics and Statistics* (1990, 1992) and *Journal of Policy Modeling* (1993).

13.2 DEFINITIONS AND EXAMPLES

The notion of cointegration is usually discussed along with such topics as nonstationarity, unit roots, and common trends. Two excellent surveying articles on trends and cointegration are those of Stock (1994) and Watson (1994). Given the large number of papers written on this subject, it seems prudent to limit our focus to a few fundamental concepts. We follow the approach used in Johansen (1996) and exemplify a couple of examples in that article to illustrate the idea. By adopting such an approach, this chapter aims at providing a useful tutorial in cointegration.

From a statistical perspective, the idea of cointegration stems from the notion of transformations. When the underlying data exhibit certain nonstationary behavior, one usually introduces transformation to render the data to stationarity. For example, the Box–Cox power transformation is often applied when nonconstant mean or variance behaviors are detected. In a time series context, Box and Tiao (1977) discuss the idea of applying canonical correlation transformations on a multivariate time series to identify stationary components.

In economics, it is often the case that although individual components of a multivariate time series $\boldsymbol{X}_t = (X_{t1}, \ldots, X_{tk})'$ appear to be nonstationary, the overall behavior of $\boldsymbol{X}_t$ may be modeled by a stationary process after a suitable transformation is applied to $\boldsymbol{X}_t$. Thus, the idea of cointegration deals with the common behavior of a multiple time series. It stems from the fact that certain linear transformations of $\boldsymbol{X}_t$ may be stationary. Engle and Granger (1987) formulate the idea of cointegration and present statistical procedures to test for cointegration for a multiple time series.

To begin, first note that nonstationarity arises when a stationary process is aggregated. For example, consider the process $X_t = \sum_{i=1}^{t} Z_i$, where Z_i is an uncorrelated sequence of random variables with mean zero and variance σ^2, usually known as a white noise sequence and denoted as $Z_t \sim \mathrm{WN}(0, \sigma^2)$. Then X_t has variance $t\sigma^2$, so that X_t is nonstationary. In this case, X_t is an example of an integrated order 1, $I(1)$ process and it is the cumulated sum of Z_t that gives rise to the nonstationarity. Specifically, we define the notion of integration for a multivariate time series $\boldsymbol{X}_t$ as follows:

Definition 13.1 *The process $\boldsymbol{X}_t = \sum_{i=0}^{\infty} \Psi_i \boldsymbol{Z}_{t-i}$ is said to be* **integrated of order zero**, *$I(0)$, if $\sum_{i=0}^{\infty} \Psi_i \neq 0$.*

To understand why the last condition is necessary in this definition, consider a simple univariate ($k = 1$) AR(1) case.

Example 13.1 *Let $X_t = \rho X_{t-1} + Z_t$, with $|\rho| < 1$ and $Z_t \sim \mathrm{WN}(0, 1)$. Clearly, $X_t = \sum_{i=0}^{\infty} \rho^i Z_{t-i}$ and $\sum_{i=0}^{\infty} \psi_i = 1/(1 - \rho) \neq 0$. According to the definition, X_t is $I(0)$. Now consider*

$$\sum_{i=1}^{t} X_i = \sum_{i=1}^{t} \rho X_{i-1} + \sum_{i=1}^{t} Z_i.$$

Rearranging terms, we have

$$(1-\rho) \sum_{i=1}^{t} X_i = \sum_{i=1}^{t} Z_i - \rho(X_t - X_0),$$

that is,

$$\sum_{i=1}^{t} X_i = (1-\rho)^{-1} \sum_{i=1}^{t} Z_i - \frac{\rho}{1-\rho} \left[\sum_{i=0}^{\infty} \rho^i (Z_{t-i} - Z_{-i}) \right].$$

This equation shows that it is the condition $0 \neq \sum_{i=0}^{\infty} \psi_i = (1-\rho)^{-1}$ that guarantees the cumulated process $\sum_{i=1}^{t} X_i$ to be nonstationary. On the other hand, the differenced process $Y_t = \Delta X_t = X_t - X_{t-1}$ is clearly stationary, but it is not necessarily $I(0)$. To see this fact, consider

$$\begin{aligned} Y_t &= X_t - X_{t-1} \\ &= \sum_{i=0}^{\infty} \rho^i Z_{t-i} - \sum_{i=0}^{\infty} \rho^i Z_{t-1-i} \\ &= Z_t + \sum_{i=1}^{\infty} \rho^i \left(1 - \frac{1}{\rho}\right) Z_{t-i} \\ &= \sum_{i=0}^{\infty} \hat{\psi}_i Z_{t-1}, \end{aligned}$$

with $\hat{\psi}_0 = 1, \hat{\psi}_i = \rho^i(1 - 1/\rho)$, $i = 1, 2, \ldots$. Direct calculation shows that $\sum \hat{\psi}_i = 0$. According to the definition, although the process $\{Y_t\}$ is stationary, it is not $I(0)$. □

Having defined the notion of $I(0)$, we can now define the notion of $I(1)$ as follows.

Definition 13.2 *A k-dimensional $\{X_t\}$ is said to be* **integrated of order 1**, *$I(1)$, if $\Delta X_t = (I - B)X_t = X_t - X_{t-1}$ is $I(0)$.*

Herein, we define $BX_t = X_{t-1}$. The difference between $I(0)$ and $I(1)$ is demonstrated by the random walk model.

Example 13.2

$$\Delta X_t = X_t - X_{t-1} = Z_t,$$

with $\{Z_t\}$ being an $I(0)$ process so that $\{X_t\}$ becomes an $I(1)$ process. Note that in this case, $X_t = X_0 + \sum_{i=1}^{t} Z_i$. □

In fact, any process $\{X_t\}$ of the form

$$X_t = C \sum_{k=1}^{t} Z_k + \sum_{k=0}^{\infty} C_k Z_{t-k}$$

is also an $I(1)$ process provided that $C \neq 0$. To check this, consider

$$\Delta X_t = X_t - X_{t-1}$$

$$= C \sum_{k=1}^{t} Z_k + \sum_{k=0}^{\infty} C_k Z_{t-k} - C \sum_{k=1}^{t-1} Z_k - \sum_{k=0}^{\infty} C_k Z_{t-1-k}$$

$$= C Z_t + \sum_{k=0}^{\infty} C_k Z_{t-k} - Z_{t-1-k}$$

$$= (C + C_0) Z_t + \sum_{k=1}^{\infty} (C_k - C_{k-1}) Z_{t-k},$$

which is $I(0)$ provided that $C + C_0 + \sum_{k=1}^{\infty} (C_k - C_{k-1}) \neq 0$ (i.e., $C \neq 0$).

To understand the real effect of cointegration on a multivariate time series, consider the following three-dimensional example.

Example 13.3 *Let*

$$X_{t1} = \sum_{i=1}^{t} Z_{i1} + Z_{t2},$$

$$X_{t2} = \frac{1}{2} \sum_{i=1}^{t} Z_{i1} + Z_{t3},$$

$$X_{t3} = Z_{t3},$$

where $(Z_{t1}, Z_{t2}, Z_{t3})' \sim \mathrm{WN}(\mathbf{0}, I_3)$ and I_3 denotes the identity matrix of order 3.

Clearly, $\boldsymbol{X}_t = (X_{t1}, X_{t2}, X_{t3})'$ is nonstationary since the first two components consist of the common term $\sum_{i=1}^{t} Z_{i1}$, called the common trend or the common nonstationary component of $\boldsymbol{X}_t$. On the other hand,

$$\Delta \boldsymbol{X}_t = (I - B) \boldsymbol{X}_t$$

$$= \boldsymbol{X}_t - \boldsymbol{X}_{t-1}$$

$$= \left(Z_{t1} + Z_{t2} - Z_{t-1,2}, \frac{1}{2} Z_{t1} + Z_{t2} - Z_{t-1,2}, Z_{t3} - Z_{t-1,3} \right)'$$

is stationary. It can easily be checked that $\Delta \boldsymbol{X}_t$ is $I(0)$, hence $\boldsymbol{X}_t$ is $I(1)$. Furthermore,

$$
\begin{aligned}
X_{t1} - 2X_{t2} &= \sum_{i=1}^{t} Z_{i1} + Z_{t2} - \sum_{i=1}^{t} Z_{i1} - 2Z_{t3} \\
&= Z_{t2} - 2Z_{t3},
\end{aligned}
$$

which is stationary as it is expressed as a finite sum of two uncorrelated white noise sequences. Although $\boldsymbol{X}_t$ is nonstationary, for $\beta' \boldsymbol{X}_t$, where $\beta = (1, -2, 0)'$ is stationary, we say that $\boldsymbol{X}_t$ is cointegrated with cointegrating vector $\beta = (1, -2, 0)'$. The part $\sum_{i=1}^{t} Z_{i1}$ is called a **common stochastic trend** *of the process $\boldsymbol{X}_t$. In this example, stationarity is attained by either differencing or by taking linear combinations.* □

As an application of this example, suppose that X_{t1} represents the wholesale price index for a specific country, X_{t2} represents a trade-weighted foreign price index, and X_{t3} represents the exchange rates between these two countries, all measured in logarithmic scales. Preceding equations indicate that the two indices are driven by the same stochastic common trend $\sum_{i=1}^{t} Z_{i1}$. The equation

$$
X_{t1} - 2X_{t2} - X_{t3} = Z_{t2} - 3Z_{t3}
$$

represents a stationary equilibrium relationship between the two indices and the exchange rate with $(1, -2, -1)'$ being an cointegrating vector. This is an example of purchasing power parity; further details can be found in Johansen (1996).

With this example as the background, we can now define cointegration precisely.

Definition 13.3 *Let $\{\boldsymbol{X}_t\}$ be an $I(1)$ process. If there exists $\beta \neq 0$ such that $\beta' \boldsymbol{X}_t$ is stationary by a suitable choice of $\beta' \boldsymbol{X}_0$, then $\boldsymbol{X}_t$ is called* **cointegrated** *and β is called the* **cointegrating vector**. *The number of linearly independent* **cointegrating vectors** *is called the* **cointegrating rank**, *and the space spanned by the cointegrating vectors is called the* **cointegrating space**.

Remark. Let $\boldsymbol{X}_t$ be $I(1)$ and let A be a matrix of full rank. Then $A\boldsymbol{X}_t$ is also $I(1)$. Further if β is the cointegrating vector, then $(A^{-1})'\beta$ is the corresponding cointegrating vector of $A\boldsymbol{X}_t$. Consequently, the notion of $I(1)$ is invariant under any nonsingular linear transformation.

13.3   ERROR CORRECTION FORM

To carry out inference for cointegrated series, Engle and Granger (1987) make use of the idea of error correction form for a multivariate time series. Again, this concept is best understood through an example.

Example 13.4 *Consider the bivariate process*

$$\Delta X_{t1} = -\alpha_1(X_{t-1,1} - 2X_{t-1,2}) + Z_{t1},$$

$$\Delta X_{t2} = Z_{t2},$$

where

$$Z_t = (Z_{t1}, Z_{t2})' \sim \text{WN}(0, I_2).$$

Several interpretations can be deduced from this example.

1. *Change in X_{t1} is related to a disequilibrium error $X_{t-1,1} - 2X_{t-1,2}$ by the adjustment coefficient α_1.*

2. *ΔX_t is $I(0)$, so that X_t is $I(1)$. To establish this, consider*

$$\Delta X_t = \begin{pmatrix} \Delta X_{t1} \\ \Delta X_{t2} \end{pmatrix}$$

$$= \begin{pmatrix} -\alpha_1 \sum_{i=0}^{\infty}(1-\alpha_1)^i(Z_{t-i-1,1} - 2Z_{t-i-1,2}) + Z_{t1} \\ Z_{t2} \end{pmatrix}$$

$$= \begin{pmatrix} 1 & 0 \\ 0 & 1 \end{pmatrix}\begin{pmatrix} Z_{t1} \\ Z_{t2} \end{pmatrix} + \begin{pmatrix} -\alpha_1 & 2\alpha_1 \\ 0 & 0 \end{pmatrix}\begin{pmatrix} Z_{t-11} \\ Z_{t-12} \end{pmatrix}$$

$$+ \begin{pmatrix} -\alpha_1(1-\alpha_1) & 2\alpha_1(1-\alpha_1) \\ 0 & 0 \end{pmatrix}\begin{pmatrix} Z_{t-2,1} \\ Z_{t-2,2} \end{pmatrix} + \cdots,$$

which implies that $\sum_{i=0}^{\infty} \Psi_i \neq 0$. So ΔX_t is $I(0)$ and X_t is $I(1)$.

3. *$X_{t1} - 2X_{t2}$ is an $\text{AR}(1)$ process. To check this, observe that*

$$X_{t1} - X_{t-1,1} = -\alpha_1(X_{t-1,1} - 2X_{t-1,2}) + Z_{t1},$$

$$X_{t2} - X_{t-1,2} = Z_{t2}.$$

Therefore,

$$X_{t1} - 2X_{t2} = (1-\alpha_1)X_{t-1,1} - (1-\alpha_1)2X_{t-1,2} + Z_{t1} - 2Z_{t2}$$

$$= (1-\alpha_1)(X_{t-1,1} - 2X_{t-1,2}) + Z_{t1} - 2Z_{t2}.$$

Consequently,

$$X_{t1} - 2X_{t2} = \sum_{i=0}^{\infty}(1-\alpha_1)^i(Z_{t-i,1} - 2Z_{t-i,2})$$

$$= \sum_{i=0}^{t-1}(1-\alpha_1)^i(Z_{t-i,1} - 2Z_{t-i,2}) + (1-\alpha_1)^t(X_{01} - 2X_{02})$$

$$= \sum_{i=0}^{\infty}(1-\alpha_1)^i(Z_{t-i,1} - 2Z_{t-i,2})$$

when the initial values $X_{01} - 2X_{02}$ are expressed in terms of past values of Z's. In particular,

$$X_{t2} = \sum_{i=1}^{t} Z_{i2} + X_{02}$$

$$X_{t1} = 2X_{t2} + \sum_{i=0}^{\infty}(1 - \alpha_1)^i(Z_{t-i,1} - 2Z_{t-i,2})$$

$$= 2\sum_{i=1}^{t} Z_{i2} + \sum_{i=0}^{\infty}(1 - \alpha_1)^i(Z_{t-i,1} - 2Z_{t-i,2}) + 2X_{02}.$$

Hence, $X_{t1} - 2X_{t2}$ is an $I(0)$ stationary AR(1) process.

4. *Accordingly, $(1, -2)'$ is the cointegrating vector, as $X_{t1} - 2X_{t2}$ is stationary and $I(0)$ and $\sum_{i=1}^{t} Z_{i2}$ is the command trend.*

5. *The idea of error correction can be interpreted as follows. Let X_{t1} represent the price of a commodity in a particular market and X_{t2} is the corresponding price of the same commodity in a different market. Suppose that the equilibrium position between the two variables is given by $X_{t1} = \gamma X_{t2}$ and change in X_{t1} is given by its deviation from this equilibrium in period $t - 1$:*

$$\Delta X_{t1} = \alpha_1(X_{t-1,1} - \gamma X_{t-1,2}) + Z_{t1}.$$

Similarly, if we apply the same interpretation to X_{t2}, we have

$$\Delta X_{t2} = \alpha_2(X_{t-1,1} - \gamma X_{t-1,2}) + Z_{t2}.$$

If X_{t1} and X_{t2} are $I(1)$, then according to the definition, ΔX_{t1} and ΔX_{t2} are $I(0)$. Further, since Z_{t1} and Z_{t2} are assumed to be stationary and $I(0)$, this implies that

$$\alpha_i(X_{t-1,1} - \gamma X_{t-1,2}) = \Delta X_{ti} - Z_{ti}$$

must be stationary (i.e., $\boldsymbol{X}_t$ is cointegrated).   □

13.4 GRANGER'S REPRESENTATION THEOREM

The ideas of Example 13.4 can be summarized succinctly by the celebrated Granger's representation theorem. Consider a k-dimensional vector autoregressive [VAR(p)] model $\boldsymbol{X}_t$ which satisfies the equation

$$\boldsymbol{X}_t = \Phi_1\boldsymbol{X}_{t-1} + \cdots + \Phi_p\boldsymbol{X}_{t-p} + \boldsymbol{Z}_t, \tag{13.1}$$

where $\{\boldsymbol{Z}_t\} \sim \mathrm{WN}(\boldsymbol{0}, I_k)$, a k-dimensional white noise process. It is well known [see, e.g., page 11 of Lütkepohl (1993)] that a sufficient condition for $\boldsymbol{X}_t$ to be stationary is that the solutions of the characteristic polynomial of (13.1) lie outside the unit circle (no unit roots allowed), that is,

$$\det(I - \Phi_1 \zeta - \cdots - \Phi_p \zeta^p) \neq 0, \quad \text{for all } |\zeta| \leq 1.$$

Now, define $\Phi(\zeta) = I - \Phi_1 \zeta - \cdots - \Phi_p \zeta^p$, $\Phi = -\Phi(1)$, and $\Gamma = -d\Phi(\zeta)/d\zeta \mid_{\zeta=1} + \Phi$. Then,

$$\Phi = -\Phi(1) = \Phi_1 + \cdots + \Phi_p - I,$$

$$\Gamma = -\left. \frac{d\Phi(\zeta)}{d\zeta} \right|_{\zeta=1} + \Phi$$

$$= \Phi_1 + 2\Phi_2 \zeta \mid_{\zeta=1} + \cdots + p\Phi_p \zeta^{p-1} \mid_{\zeta=1} + \Phi$$

$$= \Phi_1 + 2\Phi_2 + \cdots + p\Phi_p + \Phi_1 + \cdots + \Phi_p - I$$

$$= 2\Phi_1 + 3\Phi_2 + \cdots + (p+1)\Phi_p - I.$$

With these definitions, we first state a well-known condition under which a stationary $VAR(p)$ process is $I(0)$.

Theorem 13.1 *If $\boldsymbol{X}_t$ is a stationary* $\mathrm{VAR}(p)$ *process such that* $\det(\Phi(\zeta)) \neq 0$ *for all* $|\zeta| < 1$, *then $\boldsymbol{X}_t$ can be given an initial distribution such that it becomes $I(0)$ if and only if Φ is full rank. In this case, there is no unit root for $\boldsymbol{X}_t$ (stationary) and*

$$\boldsymbol{X}_t = \sum_{i=0}^{\infty} \Psi_i \boldsymbol{Z}_{t-i},$$

where $\Psi(\zeta) = \sum_{i=0}^{\infty} \Psi_i \zeta^i = (\Phi(\zeta))^{-1}$ *converges for* $|\zeta| < 1 + \delta$ *for some* $\delta > 0$.

Since $\Phi = \Phi_1 + \cdots + \Phi_p - I$, the full rank condition of Φ in this theorem holds if and only if $\det \Phi = \det(\Phi(1)) \neq 0$. In other words, the full rank condition of Φ simply means that $\det(\Phi(\zeta)) \neq 0$ for $\zeta = 1$. When this matrix is of reduced rank, we have a cointegrated system.

Definition 13.4 *Let α be any $k \times r$ matrix of rank r $(r < k)$. Define $\alpha_\perp$ to be a $k \times (k - r)$ matrix which is full rank such that $\alpha' \alpha_\perp = 0$.*

Now assume that the following condition holds for $\boldsymbol{X}_t$:

$$\det(\Phi(\zeta)) \neq 0 \quad \text{for all } |\zeta| < 1$$

(i.e., the process $\boldsymbol{X}_t$ is nonstationary since it can have unit roots). Under this assumption, we have Granger's representation theorem.

Theorem 13.2 (Granger's Representation Theorem) *Let $\boldsymbol{X}_t$ be a* $\mathrm{VAR}(p)$ *process and let* $\det(\Phi(\zeta)) \neq 0$ *for all* $|\zeta| < 1$ *(i.e., unit roots are* *allowed so that the process $\boldsymbol{X}_t$ can be nonstationary). Then $\boldsymbol{X}_t$ is $I(1)$ if and only if*

$$\Phi = \alpha\beta',$$

where α and β are $k \times r$ matrices which are of full rank r $(r < k)$, and

$$\alpha'_\perp \Gamma \beta_\perp \text{ is of full rank.}$$

In this case, $\Delta\boldsymbol{X}_t$ and $\beta'\boldsymbol{X}_t$ can be given initial distributions so that they become $I(0)$. Furthermore, under these circumstances, $\boldsymbol{X}_t$ can be expressed as

$$\boldsymbol{X}_t = \beta_\perp (\alpha'_\perp \Gamma \beta_\perp)^{-1} \alpha'_\perp \sum_{i=1}^{t} \boldsymbol{Z}_i + \Psi(B)\boldsymbol{Z}_t$$

and

$$\Delta\boldsymbol{X}_t = \alpha\beta'\boldsymbol{X}_{t-1} + \sum_{i=1}^{p-1} \Gamma_i \, \Delta\boldsymbol{X}_{t-i} + \boldsymbol{Z}_t, \tag{13.2}$$

where

$$\Gamma_i = -(\Phi_{i+1} + \cdots + \Phi_p), \quad i = 1, \ldots, p-1.$$

Note that (13.2) in this theorem is the error correction form expressed in terms of $\Delta\boldsymbol{X}_t$. Again, this theorem can best be understood in terms of a specific VAR(2) example.

Example 13.5 *Let $\boldsymbol{X}_t = \Phi_1 \boldsymbol{X}_{t-1} + \Phi_2 \boldsymbol{X}_{t-2} + \boldsymbol{Z}_t$. By differencing, we have*

$$\boldsymbol{X}_t - \boldsymbol{X}_{t-1} = -(I - \Phi_1)\boldsymbol{X}_{t-1} + \Phi_2 \boldsymbol{X}_{t-2} + \boldsymbol{Z}_t,$$

$$\begin{aligned} \Delta\boldsymbol{X}_t &= -(I - \Phi_1)\boldsymbol{X}_{t-1} + (I - \Phi_1)\boldsymbol{X}_{t-2} - (I - \Phi_1 - \Phi_2)\boldsymbol{X}_{t-2} + \boldsymbol{Z}_t \\ &= -(I - \Phi_1)(\boldsymbol{X}_{t-1} - \boldsymbol{X}_{t-2}) - (I - \Phi_1 - \Phi_2)\boldsymbol{X}_{t-2} + \boldsymbol{Z}_t \\ &= -(I - \Phi_1)\,\Delta\boldsymbol{X}_{t-1} + \Phi\boldsymbol{X}_{t-2} + \boldsymbol{Z}_t \\ &= \Phi\boldsymbol{X}_{t-2} - (I - \Phi_1)\,\Delta\boldsymbol{X}_{t-1} + \boldsymbol{Z}_t \\ &= \Phi\boldsymbol{X}_{t-1} - \Phi(\boldsymbol{X}_{t-1} - \boldsymbol{X}_{t-2}) - (I - \Phi_1)\,\Delta\boldsymbol{X}_{t-1} + \boldsymbol{Z}_t \\ &= \Phi\boldsymbol{X}_{t-1} + (\Phi_1 - I - \Phi)\,\Delta\boldsymbol{X}_{t-1} + \boldsymbol{Z}_t \\ &= \Phi\boldsymbol{X}_{t-1} + \Gamma_1\,\Delta\boldsymbol{X}_{t-1} + \boldsymbol{Z}_t, \end{aligned}$$

where $\Gamma_1 = -\Phi_2$. Note that this last expression is exactly the error correction form. Under cointegration, we have $\Phi = \alpha\beta'$ and

$$\Delta\boldsymbol{X}_t = -\alpha\beta'\boldsymbol{X}_{t-1} + \Gamma_1\,\Delta\boldsymbol{X}_{t-1} + \boldsymbol{Z}_t.$$

Therefore, the cointegrating vectors are $\alpha\beta'$ and the common trend is

$$\alpha'_\perp \sum_{i=1}^{t} \boldsymbol{Z}_i. \qquad \square$$

We now use a second example to illustrate the reduced rank structure for a VAR model.

Example 13.6

$$\Delta X_t = \begin{pmatrix} -\alpha_1 & 2\alpha_1 \\ 0 & 0 \end{pmatrix} X_{t-1} + Z_t = AX_{t-1} + Z_t.$$

In other words,

$$X_t = (I + A)X_{t-1} + Z_t = \Phi_1 X_{t-1} + Z_t,$$

where

$$\Phi_1 = I + A = \begin{pmatrix} 1 - \alpha_1 & 2\alpha_1 \\ 0 & 1 \end{pmatrix}.$$

Observe that

$$\begin{aligned}
\Phi(\zeta) &= I - \Phi_1 \zeta \\
&= I - (I + A)\zeta \\
&= \begin{pmatrix} 1 - \zeta + \alpha_1 \zeta & -2\alpha_1 \zeta \\ 0 & 1 - \zeta \end{pmatrix} \\
&= \begin{pmatrix} \alpha_1 & -2\alpha_1 \\ 0 & 0 \end{pmatrix} + (1 - \zeta) \begin{pmatrix} 1 - \alpha_1 & 2\alpha_1 \\ 0 & 1 \end{pmatrix},
\end{aligned}$$

with

$$\Phi = \Phi(1) = \begin{pmatrix} \alpha_1 & -2\alpha_1 \\ 0 & 0 \end{pmatrix},$$

and

$$\det(\Phi(\zeta)) = (1 - \zeta)(1 + \alpha_1 \zeta - \zeta).$$

This characteristic polynomial has roots $\zeta = 1$ and $\zeta = (1 - \alpha_1)^{-1}$ if $\alpha_1 \neq 1$. Therefore, for $0 < \alpha_1 < 1$, $\det \Phi(\zeta) = 0$ for $\zeta = 1$ and $\zeta = 1/(1 - \alpha_1)$, and as a result, $\det \Phi(\zeta) = 0$ for $|\zeta| \geq 1$. Clearly, Φ has reduced rank $(= 1)$ and $\Phi = \alpha\beta'$, where $\alpha = (\alpha_1, 0)', \beta = (1, -2)', \alpha_\perp = (0, 1)',$ and $\beta_\perp = (2, 1)'$. Hence,

$$\alpha_\perp' \Gamma \beta_\perp = (0, 1)(2\Phi_1 - I) \begin{pmatrix} 2 \\ 1 \end{pmatrix} = 1.$$

By letting $X_0 = 0$, the condition of Granger's representation theorem is satisfied and X_t is therefore $I(1)$. In this case,

$$\boldsymbol{X}_t = \left(\begin{array}{c} X_{t1} \\ X_{t2} \end{array} \right)$$

$$= \left(\begin{array}{c} 2\sum_{i=1}^{t} Z_{i2} + \sum_{i=0}^{\infty} (1-\alpha_1)^i (Z_{t-i,1} - Z_{t-i,2}) \\ \sum_{i=1}^{t} Z_{i2} \end{array} \right)$$

$$= \left(\begin{array}{cc} 0 & 2 \\ 0 & 1 \end{array} \right) \left(\begin{array}{c} \sum_{i=1}^{t} Z_{i1} \\ \sum_{i=2}^{t} Z_{i2} \end{array} \right) + \sum_{i=0}^{\infty} \Psi_i \boldsymbol{Z}_{t-i},$$

where

$$\Psi_i = \left(\begin{array}{cc} (1-\alpha_1)^i & -2(1-\alpha_1)^i \\ 0 & 0 \end{array} \right), \quad i = 0, 1, \ldots$$

that is,

$$X_t = \beta_\perp \alpha_\perp' \sum_{i=1}^{t} \boldsymbol{Z}_i + \sum_{i=0}^{\infty} \Psi_i \boldsymbol{Z}_{t-i}.$$

Note that the quantity $\alpha_\perp' \sum_{i=1}^{t} \boldsymbol{Z}_i$ of this equation constitutes the common trend. Substituting the definition of $\alpha_\perp = (0,1)'$ into this expression, the common trend equals to $\sum_{i=1}^{t} Z_{i2}$. Furthermore,

$$\Delta \boldsymbol{X}_t = \left(\begin{array}{cc} -\alpha_1 & 2\alpha_1 \\ 0 & 0 \end{array} \right) \boldsymbol{X}_{t-1} + \boldsymbol{Z}_t$$

$$= \left(\begin{array}{c} \alpha_1 \\ 0 \end{array} \right) (1, -2) \boldsymbol{X}_{t-1} + \boldsymbol{Z}_t$$

$$= \alpha \beta' \boldsymbol{X}_{t-1} + \boldsymbol{Z}_t.$$

In this case, the cointegrating vector is $\beta = (1, -2)'$ and $\Gamma_i = 0$ in Granger's representation theorem. $\square$

13.5   STRUCTURE OF COINTEGRATED SYSTEMS

Expressing a cointegrated system in the error correction form yields

$$\Delta \boldsymbol{X}_t = \Phi \boldsymbol{X}_{t-1} + \sum_{i=1}^{p-1} \Gamma_i \, \Delta \boldsymbol{X}_{t-1} + \boldsymbol{Z}_t,$$

$$\boldsymbol{Z}_t \sim \mathrm{N}(\boldsymbol{0}, \Sigma_{\boldsymbol{Z}}).$$

The underlying parameters consist of $(\alpha, \beta, \Gamma_1, \ldots, \Gamma_{p-1}, \Sigma_{\boldsymbol{Z}})$. There are all together $2r(k-r) + (p-1)k^2 + \frac{1}{2}k(k+1)$ parameters since α is of rank r, so it consists of $r \times (k-r)$ freely varying parameters. Herein, we shall assume that

$$\text{rank } \Phi = r, \quad 0 \leq r \leq k,$$

$$\Phi = \alpha\beta',$$

where α and β are $k \times r$ matrices of rank r. This assumption simply states that the number of unit roots in the system is $k - r$ for $0 \leq r \leq k$, so that the number of common trends ranges from k to 0. Some remarks are in order.

Remarks

1. H_0: $r = 0$; then $\Phi = 0$, which means that $\Delta\boldsymbol{X}_t$ is a stable VAR$(p-1)$ (i.e., there are k unit roots and no cointegration relationship).

2. H_k: $r = k$; then Φ is full rank, which means that $\det(\Phi(\zeta)) \neq 0$ for $|\zeta| = 1$. Hence $\boldsymbol{X}_t$ is $I(0)$ and is a stable VAR(p) satisfying

$$\boldsymbol{X}_t = \Phi_1\boldsymbol{X}_{t-1} + \cdots + \Phi_p\boldsymbol{X}_{t-p} + \boldsymbol{Z}_t.$$

3. H_r: $0 < r < k$; then Φ is of rank deficient with $\Phi = \alpha\beta'$. In this case, there exist $k - r$ unit roots and r cointegration relationships for $\boldsymbol{X}_t$ with cointegrating vectors given by the matrix β, and $\alpha'_\perp \sum_{i=1}^t \boldsymbol{Z}_i$ are the common trends.

13.6   STATISTICAL INFERENCE FOR COINTEGRATED SYSTEMS

Before we embark on estimation and testing, we need to review a few key results from multivariate analysis, one of which is the concept of canonical correlations.

13.6.1   Canonical Correlations

Let a $(p_1 + p_2)$ by 1 random vector $\boldsymbol{X}$ be partitioned as two subcomponents with dimensions p_1 and p_2, respectively, such that $\boldsymbol{X} = (\boldsymbol{X}^{(1)'}, \boldsymbol{X}^{(2)'})'$ and the corresponding covariance matrix be partitioned as

$$\Sigma = \begin{pmatrix} \Sigma_{11} & \Sigma_{12} \\ \Sigma_{21} & \Sigma_{22} \end{pmatrix}.$$

Assume that these random vectors are of mean zero and assume also that $p_1 \leq p_2$. One of the main goals in canonical correlation analysis is to find linear combinations of the subcomponents such that their correlations are biggest. In summary, the following algorithm can be used; for details, see Anderson (1984).

1. Find $U = \alpha' X^{(1)}$ and $V = \gamma' X^{(2)}$ such that $\text{corr}(U, V)$ is maximized subject to the condition that $EU^2 = EV^2 = 1$. Suppose that such linear combinations are found and let them be $\alpha^{(1)}$ and $\gamma^{(1)}$. Let $U_1 = \alpha^{(1)'} X^{(1)}$ and $V_1 = \gamma^{(1)'} X^{(2)}$ and denote the correlation between U_1 and V_1 by λ_1.

2. Find a second set of linear combinations of $X^{(1)}$ and $X^{(2)}$, U_2 and V_2, say, with $EU_2^2 = EV_2^2 = 1$ such that among all possible linear combinations which are *uncorrelated* with U_1 and V_1 from step 1, U_2 and V_2 have the largest correlation, and denote this correlation, by λ_2.

3. Continuing with this procedure, at the rth step, we have the following linear combinations $U_1 = \alpha^{(1)'} X^{(1)}$, $V_1 = \gamma^{(1)'} X^{(2)}, \ldots, U_r = \alpha^{(r)'} X^{(1)}$, $V_r = \gamma^{(r)'} X^{(2)}$ with corresponding correlations $\lambda_1 > \lambda_2 > \cdots > \lambda_r$, and $EU_i^2 = EV_i^2 = 1$, $i = 1, \ldots, r$.

Definition 13.5 *These r pairs U_r and V_r are known as the rth canonical variates, where $U_r = \gamma^{(r)'} X^{(1)}$ and $V_r = \gamma^{(r)'} X^{(2)}$, each with unit variance and uncorrelated with the first $r - 1$ pairs of canonical variates. The correlation λ_r is called the rth **canonical correlation**.*

Theorem 13.3 *The quantities $\lambda_1^2, \ldots, \lambda_{p_1}^2$ satisfy*

$$| \Sigma_{12}\Sigma_{22}^{-1}\Sigma_{21} - \lambda^2\Sigma_{11} | = 0, \tag{13.3}$$

and the vectors $\alpha^{(1)}, \ldots, \alpha^{(p_1)}$ satisfy

$$(\Sigma_{12}\Sigma_{22}^{-1}\Sigma_{21} - \lambda^2\Sigma_{11})\alpha = 0. \tag{13.4}$$

In other words, $\alpha^{(i)}$ and λ_i^2 are the eigenvector and eigenvalues of the matrix $\Sigma_{12}\Sigma_{22}^{-1}\Sigma_{21} - \Sigma_{11}$. We can give the following interpretations for canonical correlations and variates.

1. λ_1^2 can be thought of as a measure of the relative effect of $V = \gamma' X^{(2)}$ on predicting $U = \alpha' X^{(1)}$ (i.e., $\alpha' X^{(1)}$ is the linear combination of $X^{(1)}$ that can be predicted best by a linear combination of $X^{(2)}$ given by $\gamma' X^{(2)}$).

2. In the one-dimensional case, assume that U and V have mean zero and variance 1 with correlation ρ. Suppose that we want to predict U by a multiple of V, bV, say. Then the mean square error of this prediction is

$$E(U - bV)^2 = 1 - 2b\rho + b^2 = (1 - \rho^2) + (b - \rho)^2.$$

This error is minimized by taking $b = \rho$ with mean square error $(1 - \rho^2)$. Thus the greater the effect of ρ^2, the more effective V is in predicting U.

3. In practice, we need to replace the unknown covariance matrix Σ by its estimate $\hat{\Sigma}$ and then solve the sample canonical variates $\hat{\alpha}$ and sample canonical correlation $\hat{\lambda}_i^2$. These can also be shown to be the maximum likelihood estimates if X is assumed to be normally distributed. Details can be found in Anderson (1984).

13.6.2   Inference and Testing

Recall that from Granger's representation theorem, given that X_t is $I(1)$, the error correction form is

$$\Delta X_t = \Phi X_{t-1} + \sum_{i=1}^{p-1} \Gamma_i \, \Delta X_{t-i} + Z_t,$$

where $\Phi = \alpha\beta'$ is of rank $r(r < k)$, with α and β being $k \times r$ matrices. Let $Z_t \sim N(0, \Sigma_Z)$, so that X_t becomes a Gaussian time series. We now show how to construct estimators and tests for cointegrated systems by the following steps. Let the length of the underlying series be T.

1. Regress ΔX_t on ΔX_{t-i}, $i = 1, \ldots, p-1$ and form the residuals R_{0t}.

2. Regress X_{t-1} on ΔX_{t-i}, $i = 1, \ldots, p-1$ and form the residuals R_{1t}.

3. Calculate the sums

$$S_{ij} = \frac{1}{T} \sum_{t=1}^{T} R_{it} R_{jt}', \quad i, j = 0, 1.$$

Theorem 13.4 *Under the hypothesis H_r: $\Phi = \alpha\beta'$ with $\mathrm{rank}(\Phi) = r$, the maximum likelihood estimate for β is given by*

$$\hat{\beta} = (\hat{v}_1, \ldots, \hat{v}_r),$$

where each $\hat{v}_i$ is a $k \times 1$ vector obtained by solving the following eigenvalues and eigenvectors problem:

$$\mid \lambda S_{11} - S_{10} S_{00}^{-1} S_{01} \mid = 0$$

for

$$1 > \hat{\lambda}_1 > \cdots > \hat{\lambda}_r > 0,$$

with

$$(\hat{\lambda}_i S_{11} - S_{10} S_{00}^{-1} S_{01}) \hat{v}_i = 0.$$

In this case, the maximum likelihood function becomes

$$L_{\max}^{-2/T} \propto \mid S_{00} \mid \Pi_{i=1}^{r}(1 - \hat{\lambda}_i),$$

and an estimator of α is given by $\hat{\alpha} = S_{01}\hat{\beta}$.

Note that by the definition of canonical correlation, $\hat{\lambda}_i$ represents the squared canonical correlation between $\Delta \boldsymbol{X}_t$ and $\boldsymbol{X}_{t-1}$ after adjusting for the effects of $\Delta \boldsymbol{X}_{t-1}, \ldots, \Delta \boldsymbol{X}_{t-p}$. Again, the following remarks can be given to Theorem 13.4.

Remarks

1. If $r = 0$, $\Delta \boldsymbol{X}_t$ is $I(0)$, but $\boldsymbol{X}_{t-1}$ is $I(1)$, so that we expect that $\Delta \boldsymbol{X}_t$ and $\boldsymbol{X}_{t-1}$ are uncorrelated. A proof of this fact can be found in Chan and Wei (1988). This result means that all $\hat{\lambda}_i$ should be very small.

2. If $r = k$, $\Delta \boldsymbol{X}_t$ and $\boldsymbol{X}_{t-1}$ are both $I(0)$. In this case we expect significant relationships among them. As a result, the estimates of β are coming from those that correlate most with the stationary process $\Delta \boldsymbol{X}_t$ after adjusting for the effects of $\Delta \boldsymbol{X}_{t-1}, \ldots, \Delta \boldsymbol{X}_{t-p}$.

3. For r in between, we choose $\hat{\beta}$ by taking the most correlated components in the canonical correlation sense.

Turning to testing for the hypothesis of H_r (i.e., there are at most r cointegrating relationships), consider the likelihood ratio statistic as follows:

$$Q_r = -2 \ln Q(H_r \mid H_k) = -T \sum_{i=r+1}^{k} \log(1 - \hat{\lambda}_i).$$

We reject H_r in favor of H_{r+1} if Q_r is too big. The interpretation for constructing such a statistic can be argued as follows. When H_r is true, there are supposed to be r cointegrating relationships which should be captured by $\hat{\lambda}_1, \ldots, \hat{\lambda}_r$. As a consequence, $\hat{\lambda}_i$ for $i = r + 1, \ldots, k$ should then be relatively small. If on the contrary they are observed to be too big for any given series, Q_r becomes huge and there is strong evidence of rejecting H_r in favor of H_{r+1}. With this interpretation, the next step is how to compare Q_r with known distributions. In a standard multivariate context, we expect to compare Q_r with some form of a chi-squared distribution. Unfortunately, due to the nonstationarity of $\boldsymbol{X}_t$, standard asymptotic is no longer available. Instead, we need to consider the following theorem.

Theorem 13.5 *Let $\boldsymbol{X}_t$ be a VAR(p) with r cointegrating relations. Then*

$$\begin{aligned} Q_r &= -2 \log(Q(H_r \mid H_k)) \\ &= -T \sum_{i=r+1}^{k} \log(1 - \hat{\lambda}_i) \\ &\overset{\mathcal{L}}{\longrightarrow} \xi, \end{aligned}$$

where

$$\xi = \text{trace} \left\{ \int_0^1 dB(s)(B(s))' \left[\int_0^1 B(s)B'(s)ds \right]^{-1} \int_0^1 B(s)(dB(s))' \right\},$$

with $B(s)$ being the $(k-r)$-dimensional standard Brownian motion.

Notice that the limiting distribution depends only on the number $k - r$, which is the number of common trends in the underlying series $\boldsymbol{X}_t$. Tabulations of special cases of ξ can be found in a number of places, including Banerjee, Dolado, Calbraith, and Hendry (1993), Johansen (1995), Tanaka (1996), Maddala and Kim (1998), and Lütkepohl (1993). Other examples and applications of cointegrations can be found in Chapter 19 of Hamilton (1994).

13.7   EXAMPLE OF SPOT INDEX AND FUTURES

Our data set consists of 1861 daily observations of the spot index and the futures price of the Nikkei Stock Average 225 (NSA), covering the period from January 1989 through August 1996. This data set is analyzed in Lien and Tse (1999) and is stored under the file dh.dat on the Web page for this book. It can be read in SPLUS using

```
> dh.dat_read.table('dh.dat',row.names=NULL,header=T)
```

We will convert the data from daily to monthly (93 months altogether), since the SPLUS program that we introduce runs out of memory very fast with big data sets. For that reason we do

```
> spotmonth_numeric(93)
> futmonth_numeric(93)
> for(i in 0:93){
spotmonth[i+1]_mean(dh.dat$SPOT[dh.dat$INDEX==i])
futmonth[i+1]_mean(dh.dat$FUTURES[dh.dat$INDEX==i])
}
```

and two vectors with monthly data are created, spotmonth and futmonth. We convert them into regular time series and plot them, with the resulting plot shown in Figure 13.1.

```
> spotmonth_rts(spotmonth,start=c(1989,1),freq=12)
> futmonth_rts(futmonth,start=c(1989,1),freq=12)
> spotfut_ts.union(spotmonth,futmonth)
> tsplot(spotfut)
> legend(1994,35000,legend=c("Spot Index","Futures Price"),
```

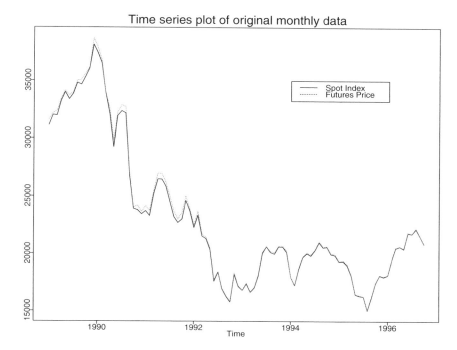

Fig. 13.1   Nikkei stock average, 1989/1 through 1996/8.

```
+   lty=1:2)
>   title('Time series plot of original monthly data')
```

Observe in Figure 13.1 how close the two series are, suggesting a possible cointegration. Besides the data, we can also download the source code that will be used to determine the cointegration factor on the Web page for this book under the file coin.q. The source code can be loaded into SPLUS using the command

```
>   source('coin.q')
```

The vector series will be modeled by a VAR model, whose order will be determined by the AIC criterion using the function coin.order, which is included in the coin.q package.

```
>   coin.order(spotfut,1,15)
[[1]]:
[1] 0

$order:
  [1]   0   1   2   3   4   5   6   7   8   9  10  11  12  13  14  15
```

$AIC:
```
            [,1]
 [1,]  30.03754
 [2,]  23.83575
 [3,]  23.79718
 [4,]  23.41154
 [5,]  23.30038
 [6,]  23.31654
 [7,]  23.25862
 [8,]  23.30175
 [9,]  23.24694
[10,]  23.33625
[11,]  23.43114
[12,]  23.44845
[13,]  23.35808
[14,]  23.19575
[15,]  22.94025
```

$HQ:
```
            [,1]
 [1,]  30.03754
 [2,]  23.87973
 [3,]  23.88569
 [4,]  23.54512
 [5,]  23.47960
 [6,]  23.54195
 [7,]  23.53082
 [8,]  23.62132
 [9,]  23.61448
[10,]  23.75237
[11,]  23.89645
[12,]  23.96360
[13,]  23.92370
[14,]  23.81248
[15,]  23.60876
```

$SC:
```
            [,1]
 [1,]  30.03754
 [2,]  23.94467
 [3,]  24.01647
 [4,]  23.74264
 [5,]  23.74479
 [6,]  23.87578
 [7,]  23.93426
 [8,]  24.09537
 [9,]  24.16019
[10,]  24.37079
[11,]  24.58867
[12,]  24.73073
[13,]  24.76690
[14,]  24.73292
[15,]  24.60767
```

The function `coin.order` accepts the cointegration rank and the maximum order of the VAR model as arguments. Notice that in our example the minimum AIC is attained at lag 15. Then use the `coinec.ml` to get the value of the likelihood ratio statistic as well as the estimate for β. For testing the hypothesis H_0: $r = r_0 = 0$, try

```
> spotfut.co0_coinec.ml(spotfut,0,15)
```

that is, we create an SPLUS object `spotfut.co0` using `coinec.ml` with arguments the bivariate series, r_0, and the VAR order. This command produces a number of outputs, as follows:

```
> names(spotfut.co0)
 [1] ""           "order"    "dY"      "Y1"     "dX"
     "M"
 [7] "R0"         "R1"       "S00"     "S10"    "S01"
     "S11"
[13] "G"          "eigM"     "eig"     "C"      "H"
     "D"
[19] "residuals"  "Zu"       "Qr"      "ar"     "Y"
     "tratio"
```

We are mostly interested in the objects C, H, and Qr, which are $\hat{\beta}$, $\hat{\alpha}$, and Q_r respectively. Those values are

```
> spotfut.co0$Qr
[1] 27.38487
> spotfut.co0$H
          [,1]
[1,] -408.4281
[2,] -420.6450
> spotfut.co0$C
          [,1]          [,2]
[1,] 0.01097465 -0.01093322
```

While for H_0: $r = r_0 = 1$ (i.e., for one unit root), we have

```
> spotfut.co1_coinec.ml(spotfut,1,15)
> spotfut.co1$Qr
[1] 0.7501444
> spotfut.co1$H
          [,1]
[1,] -408.4281
[2,] -420.6450
> spotfut.co1$C
          [,1]          [,2]
[1,] 0.01097465 -0.01093322
```

Hence, the LR statistic for testing $r = 0$ (i.e., 2 unit roots) is $Q_r = 27.38487$, while the LR statistic for testing $r = 1$ (i.e., 1 unit root) is $Q_r = 0.7501444$. Using the tabulated values of the Q_r statistic from Johansen and Juselius (1990), we select a model with one unit root. The VAR coefficients for the model can be retrieved by using

```
> spotfut.co1$ar
          [,1]      [,2]      [,3]      [,4]      [,5]      [,6]
          [,7]
[1,] -4.803346 5.514735 1.898729 -1.669563 4.579344 -4.413753
[1,]  8.778863
[2,] -5.501376 6.211529 2.124517 -1.915902 3.701566 -3.517150
[2,]  9.211856
          [,8]      [,9]     [,10]     [,11]     [,12]     [,13]
          [,14]
[1,] -8.672473 3.077767 -3.145176 -8.335677 7.885471 -5.432223
[1,]  5.400558
[2,] -9.103786 3.199721 -3.249017 -8.312419 7.847850 -5.463355
[2,]  5.428096
          [,15]     [,16]     [,17]     [,18]     [,19]     [,20]
          [,21]
[1,] -3.109619 3.313804 3.685125 -3.565751 3.726477 -3.641071
[1,]  0.5527697
[2,] -3.493974 3.679815 3.556284 -3.427241 4.110664 -4.023150
[2,]  0.5064034
          [,22]     [,23]     [,24]     [,25]     [,26]     [,27]
          [,28]
[1,] -0.5870155 0.5422025 -0.7212085 -1.612225 1.409778 3.228336
[1,] -3.174341
[2,] -0.5183685 1.0584595 -1.2518207 -1.949711 1.750407 3.396192
[2,] -3.357184
          [,29]     [,30]
[1,] -1.294167 1.600571
[2,] -1.528398 1.846917
```

which gives the final estimated model for the bivariate series:

$$\Delta \boldsymbol{X}_t = \alpha \beta' \boldsymbol{X}_{t-1} + \Gamma_1 \Delta \boldsymbol{X}_{t-1} + \cdots + \Gamma_{15} \Delta \boldsymbol{X}_{t-15} + \boldsymbol{Z}_t,$$

where

$$\alpha = \begin{pmatrix} -408.4281 \\ -420.6450 \end{pmatrix}, \quad \beta = (0.01097465, -0.01093322),$$

$$\Gamma_1 = \begin{pmatrix} -4.803346 & 5.514735 \\ -5.501376 & 6.211529 \end{pmatrix}, \dots, \Gamma_{15} = \begin{pmatrix} -1.294167 & 1.600571 \\ -1.528398 & 1.846917 \end{pmatrix}.$$

The series $W_t = 0.0110 X_{1t} + -0.0110 X_{2t}$ forms a cointegrating linear combination that should be stationary, and this series is plotted in Figure 13.2. Relative to the original two series, the series W_t certainly looks more stationary.

```
> tran.sf_as.vector(spotfut.co1$C%*%t(spotfut))
> tran.sf_rts(tran.sf,start=c(1989,1),freq=12)
> tsplot(tran.sf)
```

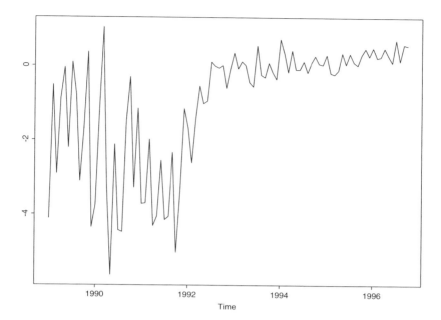

Fig. 13.2   Transformed series $W_t = 0.0110X_{1t} + -0.0110X_{2t}$.

13.8   CONCLUSIONS

As indicated in the introduction, cointegration has been an actively pursued area of research, particularly among econometricians. Owing to its simple idea, it is reasonable to expect that the concept of cointegration will be gaining its significance in other areas of social sciences. Recently, the concept of cointegration has found applications in many areas in finance. For example, Hiraki, Shiraishi, and Takezawa (1996) use cointegration and common trends to explain the term structure of offshore interest rates, and more recently, Lee, Myers, and Swaminathan (1999) used the idea of cointegration to determine the intrinsic value of the Dow Jones industrial index. Another area that is under rigorous development is combining long-memory phenomena together with cointegration, called the fractional cointegration; see, for example, Lien and Tse (1999). Owing to space limitation, we are unable to provide more descriptions about these developments. We hope this survey offers a focused and practical introduction to cointegration.

13.9   EXERCISES

1. Let $X_{t1} = \sum_{i=1}^{t} Z_{i1}$ and $X_{t2} = Z_{t2}$, where $\{Z_{i1}\}$ and $\{Z_{i2}\}$ are two independent sequences of i.i.d. random variables.

(a) Show that X_{t2} is $I(0)$.

(b) Show that X_{t1} is $I(1)$.

(c) Show that $X_t = X_{t1} + X_{t2}$ is $I(1)$.

2. If X_t is $I(d_1)$ and Y_t is $I(d_2)$ with $d_1 \neq d_2$, show that $X_t + Y_t$ is $I(d)$, where $d = \max\{d_1, d_2\}$.

3. In Example 13.3, show that the vector $(0, 0, 1)$ is also a cointegrating vector.

4. Consider the equations

$$\Delta X_{t1} = -\frac{1}{2}(X_{t-1,1} - X_{t-1,2}) + Z_{t1},$$

$$\Delta X_{t2} = \frac{1}{4}(X_{t-1,1} - X_{t-1,2}) + Z_{t2},$$

where $\{Z_{t1}\}$ and $\{Z_{t2}\}$ are two independent sequences of i.i.d. random variables.

(a) Solve these equations for X_{t1} and X_{t2} in terms of Z_{t1} and Z_{t2}.

(b) Find an equation for $X_{t1} - X_{t2}$ and another equation for $X_{t1} + 2X_{t2}$,

(c) Show by direct calculations that $X_{t1} - X_{t2}$ is stationary, but the processes X_{t1} and X_{t2} are nonstationary.

(d) What happens if $\frac{1}{4}$ is replaced by $-\frac{1}{4}$?

References

1. Abraham, B., and Ledolter, J. (1983). *Statistical Methods for Forecasting.* Wiley, New York.

2. Anderson, T. W. (1984). *An Introduction to Multivariate Statistical Analysis*, 2nd ed. Wiley, New York.

3. Aoki, M. (1990). *State Space Modeling of Time Series*, 2nd ed. Springer-Verlag, New York.

4. Babbs, S. H., and Nowman, K. B. (1999). Kalman filtering of generalized Vasicek term structure models. *Journal of Financial and Quantitative Analysis* **34**, 115–130.

5. Banerjee, A., Dolado, J. J., Galbraith, J. W., and Hendry, D. F. (1993). *Cointegration, Error Correction, and the Econometric Analysis of Nonstationary Data.* Oxford University Press, Oxford.

6. Billingsley, P. (1999). *Convergence of Probability Measures*, 2nd ed. Wiley, New York.

7. Bloomfield, P. (2000). *Fourier Analysis of Time Series: An Introduction*, 2nd ed. Wiley, New York.

8. Box, G. E. P., and Tiao, G. C. (1977). A canonical analysis of multiple time series. *Biometrika* **64**, 355–365.

9. Box, G. E. P., Jenkins, G. M., and Reinsel, G. C. (1994). *Time Series Analysis: Forecasting and Control*, 3rd ed. Prentice Hall, Upper Saddle River, N.J.

10. Brockwell, P. J., and Davis, R. A. (1991). *Time Series: Theory and Methods*, 2nd ed. Springer-Verlag, New York.

11. Brockwell, P. J., and Davis, R. A. (1996). *Introduction to Time Series and Forecasting.* Springer-Verlag, New York.

12. Campbell, J. Y., Lo, A. W., and MacKinlay, A. C. (1997). *The Econometrics of Financial Markets.* Princeton University Press, Princeton, N.J.

13. Chan, N. H., and Petris, G. (2000). Recent developments in heteroskedastic financial series, in Chan, W. S., Li, W. K., and Tong, H. (eds.), *Statistics and Fincance: An Interface.* Imperial College Press, London, pp. 169–184.

14. Chan, N. H., and Tsay, R. S. (1996). On the use of canonical correlation analysis in testing common trends, in Lee, J. C., Johnson, W. O., and Zellner, A. (eds.), *Modelling and Prediction: Honoring S. Geisser.* Springer-Verlag, New York, pp. 364–377.

15. Chan, N. H., and Wei, C. Z. (1988). Limiting distributions of least squares estimates of unstable autoregressive processes. *Annals of Statistics* **16**, 367–401.

16. Chatfield, C. (1996). *The Analysis of Time Series,* 5th ed. Chapman & Hall, New York.

17. Diebold, F. X., and Nerlove, M. (1989). The dynamics of exchange rate volatility: a multivariate latent factor. *Journal of Applied Econometrics* **4**, 1–21.

18. Diggle, P. J. (1990). *Time Series: A Biostatistical Introduction.* Clarendon Press, Oxford.

19. Engle, R. F., and Granger, C. W. J. (1987). Cointegration and error correction: representation, estimation, and testing. *Econometrica* **55**, 251–276.

20. Engle, R. F., and Granger, C. W. J. (1991). *Long-Run Economic Relations: Readings and Cointegration.* Oxford University Press, Oxford.

21. Engle, R. F., and Kroner, K. (1995). Multivariate simultaneous generalized ARCH. *Econometric Theory* **11**, 122–150.

22. Feller, W. (1968). *An Introduction to Probability Theory and Its Applications*, 3rd ed. Wiley, New York.

23. Fuller, W. A. (1996). *Introduction to Statistical Time Series*, 2nd ed. Wiley, New York.

24. Ghysels, E., Harvey, A. C., and Renault, E. (1996). Stochastic volatility. In Maddala, G. S., and Rao, C. R. (eds.), *Handbook of Statistics*, Vol. 14. Elsevier Science, Amsterdam, The Netherlands.

25. Gouriéroux, C. (1997). *ARCH Models and Financial Applications*. Springer-Verlag, New York.

26. Hamilton, J. D. (1994). *Time Series Analysis*. Princeton University Press, Princeton, N.J.

27. Hannan, E. J. (1970). *Multiple Time Series*. Wiley, New York.

28. Hannan, E. J., and Deistler, M. (1988). *The Statistical Theory of Linear Systems*. Wiley, New York.

29. Harvey, A. C. (1993). *Time Series Models*, 2nd ed. Prentice Hall, Upper Saddle River, N.J.

30. Harvey, A., Ruiz, E., and Shephard, N. (1994). Multivariate stochastic variance models. *Review of Economic Studies* **61**, 247–264.

31. Hiraki, T., Shiraishi, N., and Takezawa, N. (1996). Cointegration, common factors, and term-structure of yen offshore interest rates. *Journal of Fixed Income*, December, 69–75.

32. Jacquier, E., Polson, N., and Rossi, P. (1994). Bayesian analysis of stochastic volatility models. *Journal of Business and Economic Statistics* **12**, 371–389.

33. Johansen, S. (1995). *Likelihood-Based Inference in Cointegrated Vector Autoregressive Models*. Oxford University Press, Oxford.

34. Johansen, S. (1996). Likelihood-based inference for cointegration of some nonstationary time series. In Cox, D. R., Hinkley, D. V., and Barndorff-Nielsen, O. E. (eds.), *Time Series Models: In Econometrics, Finance and Other Fields*. Chapman & Hall, New York.

35. Johansen, S., and Juselius, K. (1990). Maximum likelihood estimation and inference on cointegration: with applications to the demand for money. *Oxford Bulletin of Economics and Statistics* **52**, 169–210.

36. Kailath, T. (1980). *Linear Systems*. Prentice Hall, Upper Saddle River, N.J.

37. Kao, D. L., and Shumaker, R. D. (1999). Equity style timing. *Financial Analysts Journal*, February, 37–48.

38. Kendall, M., and Ord, J. K. (1990). *Time Series*, 3rd ed. Oxford University Press, New York.

39. Kim, S., Shephard, N., and Chib, S. (1998). Stochastic volatility: likelihood inference and comparison with ARCH models. *Review of Economic Studies* **65**, 361–393.

40. Koopmans, L. H. (1995). *The Spectral Analysis of Time Series*. Academic Press, San Diego, Calif.

41. Krause, A., and Olson, M. (1997). *The Basics of S and S-Plus*. Springer-Verlag, New York.

42. Lang, S. (1965). *Algebra*. Addison-Wesley, Reading, Mass. p.105.

43. Lee, C. M. C., Myers, J., and Swaminathan, B. (1999). What is the intrinsic value of the Dow? *Journal of Finance* **54**, 1693–1741.

44. Lien, D., and Tse, Y. K. (1999). Forecasting the Nikkei spot index with fractional cointegration. *Journal of Forecasting* **18**, 259–273.

45. Lütkepohl, H. (1993). *Introduction to Multiple Time Series Analysis*, 2nd ed. Springer-Verlag, New York.

46. Maddala, G. S., and Kim, I. M. (1998). *Unit Roots, Cointegration, and Structural Change*. Cambridge University Press, Cambridge.

47. Melino, A., and Turnbull, S. (1990). Pricing foreign currency options with stochastic volatility. *Journal of Econometrics* **45**, 239–265.

48. Nelson, D. B. (1990). Stationarity and persistence in the GARCH(1,1) model. *Econometric Theory* **6**, 318–334.

49. Øksendal, B. (1998). *Stochastic Differential Equations: An Introduction with Applications*, 5th ed. Springer-Verlag, New York.

50. Park, J. Y. (1992). Canonical cointegrating regressions. *Econometrica* **60**, 119–143.

51. Phillips, P. C. B. (1991). Optimal inference in cointegrated systems. *Econometrica* **59**, 283–306.

52. Priestley, M. B. (1981). *Spectral Analysis and Time Series*. Academic Press, San Diego, Calif.

53. Reinsel, G. C., and Ahn, S. K. (1992). Vector AR models with unit roots and reduced rank structure: estimation, likelihood ratio test, and forecasting. *Journal of Time Series Analysis* **13**, 353–375.

54. Sandmann, G., and Koopman, S. J. (1998). Estimation of stochastic volatility models via Monte Carlo maximum likelihood. *Journal of Econometrics* **87**, 271–301.

55. Shephard, N. (1996). Statistical aspects of ARCH and stochastic volatility. In Cox, D. R., Hinkley, D. V., and Barndorff-Nielsen, O. E. (eds.), *Time Series Models: In Econometrics, Finance and Other Fields*. Chapman & Hall, New York.

56. Shumway, R. H., and Stoffer, D. (2000). *Time Series Analysis and Its Applications*. Springer-Verlag, New York.

57. Stock, J. H. (1994). Unit roots, structural breaks and trends. In Engle, R. F., and McFadden, D. L. (eds.), *Handbook of Econometrics*, Vol. IV. Elsevier, Amsterdam, The Netherlands.

58. Stock, J. H., and Watson, M. W. (1988). Testing for common trends. *Journal of the American Statistical Association* **83**, 1097–1107.

59. Tanaka, K. (1996). *Time Series Analysis: Nonstationary and Noninvertible Distribution Theory*. Wiley, New York.

60. Taylor, S. J. (1994). Modeling stochastic volatility: a review and comparative study. *Mathematics of Finance* **4**, 183–204.

61. Venables, W. N., and Ripley, B. D. (1999). *Modern Applied Statistics with S-Plus*, 3rd ed. Springer-Verlag, New York.

62. Watson, M. W. (1994). Vector autoregressions and cointegration. In Engle, R. F., and McFadden, D. L. (eds.), *Handbook of Econometrics*, Vol. IV. Elsevier, Amsterdam, The Netherlands.

63. Weisberg, S. (1985). *Applied Linear Regression*, 2nd ed. Wiley, New York.

64. West, M., and Harrison, J. (1997). *Bayesian Forecasting and Dynamic Models*, 2nd ed. Springer-Verlag, New York.

Index

Additive, 8
Alternating, 19
ARMAX, 138
Asymptotic normally distributed, 19
Asymptotically stationary, 28
Autocorrelation function, 17
  sample autocorrelation function, 18
Autocovariance function, 17
  sample autocovariance function, 18
Autoregressive integrated moving average
    model, 33
  ARIMA, 33
Autoregressive moving average model, 32
  AR model, 25
  AR(2), 30
  ARMA, 32
  SARIMA, 35

Backshift operator, 7
Bandwidth, 86
Bartlett (triangular) window, 88
BEKK model, 156
Bivariate time series, 118
Box–Cox transformation, 94

Canonical correlation, 185
Canonical observable representation,
    152
Causal, 27, 28
  condition, 139
Characteristic polynomial, 29
  reverse, 126

Circular scatter plot, 166
Cointegrated, 177
  cointegrating rank, 177
  cointegrating space, 177
  cointegrating vector, 177
Common stochastic trend, 177
Conditional least squares, 43, 47
  mean, 70
  variance, 70
Constant correlation, 157
Correlogram, 19
Curse of dimensionality, 117
Cycle, 79

Damped sine wave, 31
Decibel, 85
Diagnostic plot, 60
  diagnostic statistics, 66
Dickey–Fuller statistic, 97
Differencing, 7
Dirichlet kernel, 88
Durbin–Levinson, 41

Ensemble, 16
Equity-style timing, 3
Equivalent degrees of freedom, 88
Ergordic, 18
Error correction form, 177
Exact likelihood method, 46
Exponential smoothing, 70
Exponential smoothing technique,
    7

Filter, 5
  high-pass, 7
  low-pass, 7
  moving average, 5
  Spencer, 6
Finite-dimensional distribution, 16
Fitted residuals, 60
Forecasts, 69
  ex ante, 69
  ex post, 69
  forecast error, 70
Fourier frequency, 84
FPE (final prediction error), 49
Frequency, 79
Functional central limit theorem, 96

Gauss–Newton method, 43
Granger's representation theorem, 180

Hadamard product, 156
Heavy-tailed, 102
Heterogeneity, 101
Heteroskedastic, 101
  ARCH, 103
  autoregressive conditional
      heteroskedastic, 103
  GARCH, 105
  IGRACH, 106
  mgarch, 159
  mgarch.predict, 167
  multivariate GARCH, 154
Holt–Winters, 71
Horizon, 69

Identifiability, 117
Information index, 50
  Akaike's information criterion (AIC),
      52
  Bayesian information criterion (BIC),
      52
  Kullback–Leibler information
      index, 50
Integrated of order 1, 175
  I(1), 174
Integrated of order zero, 174
  I(0), 175
Invariance principle, 96
Invertible, 25
Itô's rule, 97

Kalman, 140
  filtering, 140
  prediction, 140
  smoothing, 140
Kolmogorov's Consistency Theorem,
    16

Lag window, 87
Lag window spectral density estimator,
    87
Lagrange multiplier test, 109
Latent, 142
Lead time, 69
Least squares, 5
Likelihood ratio statistic, 187
Logarithmic transformation, 94

Macroscopic, 5
Maximum likelihood, 44
Maximum likelihood estimate, 45
Microscopic, 5
Moment estimates, 39
Moving average method, 8
Moving average model, 24
  MA(1), 24
Multiplicative, 8
Multivariate, 117

Non-Gaussian, 143
Nonlinear, 143
Nonstationary time series, 93

Observation equation, 137

Partial autocorrelation coefficient
    (PACF), 47
Period, 79
Periodogram, 83
Persistent, 106
PMLE, 108
Portmanteau statistics, 53
  p-values of Portmanteau statistics, 64
  Portmanteau tests for residuals, 109
Positive semidefinite property, 119
Projection equation, 47
Pseudoperiodic behavior, 31
Purchasing power parity, 177

QMLE, 108

Random walk model, 34, 96
Random walk plus noise, 139
Realization, 16
Rectangular window, 88
Return, 34

Sample function, 16
Sample path, 16
Seasonal component, 8
Seasonal differencing, 8
Seasonal part, 5
Short-memory behavior, 17
Signal plus noise model, 70

Signal-to-noise ratio, 139
Simulations, 98
Single-factor model, 156
Smoother, 5
Spectral analysis, 79
    spectral density function, 80
    spectral distribution function, 80
    spectral representation theorems, 80
Spline, 8
SPLUS, 4
Square-root transformation, 94
Stable, 124, 139
Standard Brownian motion, 96
Standardized residuals, 64
State equation, 137
State space model, 137
Stationarity, 17
Stationary, 16
    differenced stationary, 95
    second-order stationary, 17
    strictly stationary, 16
    trend stationary, 95
    weakly stationary, 17
    wide-sense stationary, 17
Stochastic process, 15
Stochastic volatility models, 142
Structural model, 139
Stylized-facts, 102
Subdiagonal plots, 128
Superdiagonal plots, 128

Term-structure models, 76
    Vasicek term-structure, 144
Three-point moving average, 6
Time-domain, 39
Time-invariant systems, 138
Trend, 5
    time trend, 5
    trend stationary, 95
Two-way ANOVA, 36

Unconditional least squares, 46
Unit root tests, 96

VAR, 124
Variance-stabilizing transformation, 94
VARMA, 117
Vech, 154
Vector-diagonal, 155
Vector-valued, 117
Volatility, 101

Web page for this book, 4
Weights, 6
White noise, 83
White noise sequence, 17
Whitening matrix transformation, 163
Wiener process, 98

Yule–Walker, 30
    estimates, 41

WILEY SERIES IN PROBABILITY AND STATISTICS
ESTABLISHED BY WALTER A. SHEWHART AND SAMUEL S. WILKS

Editors
*David J. Balding, Peter Bloomfield, Noel A. C. Cressie, Nicholas I. Fisher,
Iain M. Johnstone, J. B. Kadane, Louise M. Ryan, David W. Scott,
Adrian F. M. Smith, Jozef L. Teugels*
Editors Emeriti: *Vic Barnett, J. Stuart Hunter, David G. Kendall*

The **Wiley Series in Probability and Statistics** is well established and authoritative. It covers many topics of current research interest in both pure and applied statistics and probability theory. Written by leading statisticians and institutions, the titles span both state-of-the-art developments in the field and classical methods.

Reflecting the wide range of current research in statistics, the series encompasses applied, methodological and theoretical statistics, ranging from applications and new techniques made possible by advances in computerized practice to rigorous treatment of theoretical approaches.

This series provides essential and invaluable reading for all statisticians, whether in academia, industry, government, or research.

ABRAHAM and LEDOLTER · Statistical Methods for Forecasting
AGRESTI · Analysis of Ordinal Categorical Data
AGRESTI · An Introduction to Categorical Data Analysis
AGRESTI · Categorical Data Analysis
ANDĚL · Mathematics of Chance
ANDERSON · An Introduction to Multivariate Statistical Analysis, *Second Edition*
*ANDERSON · The Statistical Analysis of Time Series
ANDERSON, AUQUIER, HAUCK, OAKES, VANDAELE, and WEISBERG ·
    Statistical Methods for Comparative Studies
ANDERSON and LOYNES · The Teaching of Practical Statistics
ARMITAGE and DAVID (editors) · Advances in Biometry
ARNOLD, BALAKRISHNAN, and NAGARAJA · Records
*ARTHANARI and DODGE · Mathematical Programming in Statistics
*BAILEY · The Elements of Stochastic Processes with Applications to the Natural
    Sciences
BALAKRISHNAN and KOUTRAS · Runs and Scans with Applications
BARNETT · Comparative Statistical Inference, *Third Edition*
BARNETT and LEWIS · Outliers in Statistical Data, *Third Edition*
BARTOSZYNSKI and NIEWIADOMSKA-BUGAJ · Probability and Statistical Inference
BASILEVSKY · Statistical Factor Analysis and Related Methods: Theory and
    Applications
BASU and RIGDON · Statistical Methods for the Reliability of Repairable Systems
BATES and WATTS · Nonlinear Regression Analysis and Its Applications
BECHHOFER, SANTNER, and GOLDSMAN · Design and Analysis of Experiments for
    Statistical Selection, Screening, and Multiple Comparisons
BELSLEY · Conditioning Diagnostics: Collinearity and Weak Data in Regression
BELSLEY, KUH, and WELSCH · Regression Diagnostics: Identifying Influential
    Data and Sources of Collinearity
BENDAT and PIERSOL · Random Data: Analysis and Measurement Procedures,
    Third Edition

*Now available in a lower priced paperback edition in the Wiley Classics Library.

BERRY, CHALONER, and GEWEKE · Bayesian Analysis in Statistics and Econometrics: Essays in Honor of Arnold Zellner

BERNARDO and SMITH · Bayesian Theory

BHAT · Elements of Applied Stochastic Processes, *Second Edition*

BHATTACHARYA and JOHNSON · Statistical Concepts and Methods

BHATTACHARYA and WAYMIRE · Stochastic Processes with Applications

BILLINGSLEY · Convergence of Probability Measures, *Second Edition*

BILLINGSLEY · Probability and Measure, *Third Edition*

BIRKES and DODGE · Alternative Methods of Regression

BLISCHKE AND MURTHY · Reliability: Modeling, Prediction, and Optimization

BLOOMFIELD · Fourier Analysis of Time Series: An Introduction, *Second Edition*

BOLLEN · Structural Equations with Latent Variables

BOROVKOV · Ergodicity and Stability of Stochastic Processes

BOULEAU · Numerical Methods for Stochastic Processes

BOX · Bayesian Inference in Statistical Analysis

BOX · R. A. Fisher, the Life of a Scientist

BOX and DRAPER · Empirical Model-Building and Response Surfaces

*BOX and DRAPER · Evolutionary Operation: A Statistical Method for Process Improvement

BOX, HUNTER, and HUNTER · Statistics for Experimenters: An Introduction to Design, Data Analysis, and Model Building

BOX and LUCEÑO · Statistical Control by Monitoring and Feedback Adjustment

BRANDIMARTE · Numerical Methods in Finance: A MATLAB-Based Introduction

BROWN and HOLLANDER · Statistics: A Biomedical Introduction

BRUNNER, DOMHOF, and LANGER · Nonparametric Analysis of Longitudinal Data in Factorial Experiments

BUCKLEW · Large Deviation Techniques in Decision, Simulation, and Estimation

CAIROLI and DALANG · Sequential Stochastic Optimization

CHAN · Time Series: Applications to Finance

CHATTERJEE and HADI · Sensitivity Analysis in Linear Regression

CHATTERJEE and PRICE · Regression Analysis by Example, *Third Edition*

CHERNICK · Bootstrap Methods: A Practitioner's Guide

CHILÈS and DELFINER · Geostatistics: Modeling Spatial Uncertainty

CHOW and LIU · Design and Analysis of Clinical Trials: Concepts and Methodologies

CLARKE and DISNEY · Probability and Random Processes: A First Course with Applications, *Second Edition*

*COCHRAN and COX · Experimental Designs, *Second Edition*

CONGDON · Bayesian Statistical Modelling

CONOVER · Practical Nonparametric Statistics, *Second Edition*

COOK · Regression Graphics

COOK and WEISBERG · Applied Regression Including Computing and Graphics

COOK and WEISBERG · An Introduction to Regression Graphics

CORNELL · Experiments with Mixtures, Designs, Models, and the Analysis of Mixture Data, *Third Edition*

COVER and THOMAS · Elements of Information Theory

COX · A Handbook of Introductory Statistical Methods

*COX · Planning of Experiments

CRESSIE · Statistics for Spatial Data, *Revised Edition*

CSÖRGŐ and HORVÁTH · Limit Theorems in Change Point Analysis

DANIEL · Applications of Statistics to Industrial Experimentation

DANIEL · Biostatistics: A Foundation for Analysis in the Health Sciences, *Sixth Edition*

*DANIEL · Fitting Equations to Data: Computer Analysis of Multifactor Data, *Second Edition*

*Now available in a lower priced paperback edition in the Wiley Classics Library.

DAVID · Order Statistics, *Second Edition*

*DEGROOT, FIENBERG, and KADANE · Statistics and the Law

DEL CASTILLO · Statistical Process Adjustment for Quality Control

DETTE and STUDDEN · The Theory of Canonical Moments with Applications in Statistics, Probability, and Analysis

DEY and MUKERJEE · Fractional Factorial Plans

DILLON and GOLDSTEIN · Multivariate Analysis: Methods and Applications

DODGE · Alternative Methods of Regression

*DODGE and ROMIG · Sampling Inspection Tables, *Second Edition*

*DOOB · Stochastic Processes

DOWDY and WEARDEN · Statistics for Research, *Second Edition*

DRAPER and SMITH · Applied Regression Analysis, *Third Edition*

DRYDEN and MARDIA · Statistical Shape Analysis

DUDEWICZ and MISHRA · Modern Mathematical Statistics

DUNN and CLARK · Applied Statistics: Analysis of Variance and Regression, *Second Edition*

DUNN and CLARK · Basic Statistics: A Primer for the Biomedical Sciences, *Third Edition*

DUPUIS and ELLIS · A Weak Convergence Approach to the Theory of Large Deviations

*ELANDT-JOHNSON and JOHNSON · Survival Models and Data Analysis

ETHIER and KURTZ · Markov Processes: Characterization and Convergence

EVANS, HASTINGS, and PEACOCK · Statistical Distributions, *Third Edition*

FELLER · An Introduction to Probability Theory and Its Applications, Volume I, *Third Edition,* Revised; Volume II, *Second Edition*

FISHER and VAN BELLE · Biostatistics: A Methodology for the Health Sciences

*FLEISS · The Design and Analysis of Clinical Experiments

FLEISS · Statistical Methods for Rates and Proportions, *Second Edition*

FLEMING and HARRINGTON · Counting Processes and Survival Analysis

FULLER · Introduction to Statistical Time Series, *Second Edition*

FULLER · Measurement Error Models

GALLANT · Nonlinear Statistical Models

GHOSH, MUKHOPADHYAY, and SEN · Sequential Estimation

GIFI · Nonlinear Multivariate Analysis

GLASSERMAN and YAO · Monotone Structure in Discrete-Event Systems

GNANADESIKAN · Methods for Statistical Data Analysis of Multivariate Observations, *Second Edition*

GOLDSTEIN and LEWIS · Assessment: Problems, Development, and Statistical Issues

GREENWOOD and NIKULIN · A Guide to Chi-Squared Testing

GROSS and HARRIS · Fundamentals of Queueing Theory, *Third Edition*

*HAHN · Statistical Models in Engineering

HAHN and MEEKER · Statistical Intervals: A Guide for Practitioners

HALD · A History of Probability and Statistics and their Applications Before 1750

HALD · A History of Mathematical Statistics from 1750 to 1930

HAMPEL · Robust Statistics: The Approach Based on Influence Functions

HANNAN and DEISTLER · The Statistical Theory of Linear Systems

HEIBERGER · Computation for the Analysis of Designed Experiments

HEDAYAT and SINHA · Design and Inference in Finite Population Sampling

HELLER · MACSYMA for Statisticians

HINKELMAN and KEMPTHORNE: · Design and Analysis of Experiments, Volume 1: Introduction to Experimental Design

HOAGLIN, MOSTELLER, and TUKEY · Exploratory Approach to Analysis of Variance

HOAGLIN, MOSTELLER, and TUKEY · Exploring Data Tables, Trends and Shapes

---

*Now available in a lower priced paperback edition in the Wiley Classics Library.

*HOAGLIN, MOSTELLER, and TUKEY · Understanding Robust and Exploratory
   Data Analysis
HOCHBERG and TAMHANE · Multiple Comparison Procedures
HOCKING · Methods and Applications of Linear Models: Regression and the Analysis
   of Variables
HOEL · Introduction to Mathematical Statistics, *Fifth Edition*
HOGG and KLUGMAN · Loss Distributions
HOLLANDER and WOLFE · Nonparametric Statistical Methods, *Second Edition*
HOSMER and LEMESHOW · Applied Logistic Regression, *Second Edition*
HOSMER and LEMESHOW · Applied Survival Analysis: Regression Modeling of
   Time to Event Data
HØYLAND and RAUSAND · System Reliability Theory: Models and Statistical Methods
HUBER · Robust Statistics
HUBERTY · Applied Discriminant Analysis
HUNT and KENNEDY · Financial Derivatives in Theory and Practice
HUSKOVA, BERAN, and DUPAC · Collected Works of Jaroslav Hajek—
   with Commentary
IMAN and CONOVER · A Modern Approach to Statistics
JACKSON · A User's Guide to Principle Components
JOHN · Statistical Methods in Engineering and Quality Assurance
JOHNSON · Multivariate Statistical Simulation
JOHNSON and BALAKRISHNAN · Advances in the Theory and Practice of Statistics: A
   Volume in Honor of Samuel Kotz
JUDGE, GRIFFITHS, HILL, LÜTKEPOHL, and LEE · The Theory and Practice of
   Econometrics, *Second Edition*
JOHNSON and KOTZ · Distributions in Statistics
JOHNSON and KOTZ (editors) · Leading Personalities in Statistical Sciences: From the
   Seventeenth Century to the Present
JOHNSON, KOTZ, and BALAKRISHNAN · Continuous Univariate Distributions,
   Volume 1, *Second Edition*
JOHNSON, KOTZ, and BALAKRISHNAN · Continuous Univariate Distributions,
   Volume 2, *Second Edition*
JOHNSON, KOTZ, and BALAKRISHNAN · Discrete Multivariate Distributions
JOHNSON, KOTZ, and KEMP · Univariate Discrete Distributions, *Second Edition*
JUREČKOVÁ and SEN · Robust Statistical Procedures: Aymptotics and Interrelations
JUREK and MASON · Operator-Limit Distributions in Probability Theory
KADANE · Bayesian Methods and Ethics in a Clinical Trial Design
KADANE AND SCHUM · A Probabilistic Analysis of the Sacco and Vanzetti Evidence
KALBFLEISCH and PRENTICE · The Statistical Analysis of Failure Time Data
KASS and VOS · Geometrical Foundations of Asymptotic Inference
KAUFMAN and ROUSSEEUW · Finding Groups in Data: An Introduction to Cluster
   Analysis
KENDALL, BARDEN, CARNE, and LE · Shape and Shape Theory
KHURI · Advanced Calculus with Applications in Statistics
KHURI, MATHEW, and SINHA · Statistical Tests for Mixed Linear Models
KLUGMAN, PANJER, and WILLMOT · Loss Models: From Data to Decisions
KLUGMAN, PANJER, and WILLMOT · Solutions Manual to Accompany Loss Models:
   From Data to Decisions
KOTZ, BALAKRISHNAN, and JOHNSON · Continuous Multivariate Distributions,
   Volume 1, *Second Edition*
KOTZ and JOHNSON (editors) · Encyclopedia of Statistical Sciences: Volumes 1 to 9
   with Index
KOTZ and JOHNSON (editors) · Encyclopedia of Statistical Sciences: Supplement
   Volume

*Now available in a lower priced paperback edition in the Wiley Classics Library.

KOTZ, READ, and BANKS (editors) · Encyclopedia of Statistical Sciences: Update Volume 1

KOTZ, READ, and BANKS (editors) · Encyclopedia of Statistical Sciences: Update Volume 2

KOVALENKO, KUZNETZOV, and PEGG · Mathematical Theory of Reliability of Time-Dependent Systems with Practical Applications

LACHIN · Biostatistical Methods: The Assessment of Relative Risks

LAD · Operational Subjective Statistical Methods: A Mathematical, Philosophical, and Historical Introduction

LAMPERTI · Probability: A Survey of the Mathematical Theory, *Second Edition*

LANGE, RYAN, BILLARD, BRILLINGER, CONQUEST, and GREENHOUSE · Case Studies in Biometry

LARSON · Introduction to Probability Theory and Statistical Inference, *Third Edition*

LAWLESS · Statistical Models and Methods for Lifetime Data

LAWSON · Statistical Methods in Spatial Epidemiology

LE · Applied Categorical Data Analysis

LE · Applied Survival Analysis

LEE · Statistical Methods for Survival Data Analysis, *Second Edition*

LePAGE and BILLARD · Exploring the Limits of Bootstrap

LEYLAND and GOLDSTEIN (editors) · Multilevel Modelling of Health Statistics

LIAO · Statistical Group Comparison

LINDVALL · Lectures on the Coupling Method

LINHART and ZUCCHINI · Model Selection

LITTLE and RUBIN · Statistical Analysis with Missing Data

LLOYD · The Statistical Analysis of Categorical Data

MAGNUS and NEUDECKER · Matrix Differential Calculus with Applications in Statistics and Econometrics, *Revised Edition*

MALLER and ZHOU · Survival Analysis with Long Term Survivors

MALLOWS · Design, Data, and Analysis by Some Friends of Cuthbert Daniel

MANN, SCHAFER, and SINGPURWALLA · Methods for Statistical Analysis of Reliability and Life Data

MANTON, WOODBURY, and TOLLEY · Statistical Applications Using Fuzzy Sets

MARDIA and JUPP · Directional Statistics

MASON, GUNST, and HESS · Statistical Design and Analysis of Experiments with Applications to Engineering and Science

McCULLOCH and SEARLE · Generalized, Linear, and Mixed Models

McFADDEN · Management of Data in Clinical Trials

McLACHLAN · Discriminant Analysis and Statistical Pattern Recognition

McLACHLAN and KRISHNAN · The EM Algorithm and Extensions

McLACHLAN and PEEL · Finite Mixture Models

McNEIL · Epidemiological Research Methods

MEEKER and ESCOBAR · Statistical Methods for Reliability Data

MEERSCHAERT and SCHEFFLER · Limit Distributions for Sums of Independent Random Vectors: Heavy Tails in Theory and Practice

*MILLER · Survival Analysis, *Second Edition*

MONTGOMERY, PECK, and VINING · Introduction to Linear Regression Analysis, *Third Edition*

MORGENTHALER and TUKEY · Configural Polysampling: A Route to Practical Robustness

MUIRHEAD · Aspects of Multivariate Statistical Theory

MURRAY · X-STAT 2.0 Statistical Experimentation, Design Data Analysis, and Nonlinear Optimization

MYERS and MONTGOMERY · Response Surface Methodology: Process and Product Optimization Using Designed Experiments, *Second Edition*

MYERS, MONTGOMERY, and VINING · Generalized Linear Models. With Applications in Engineering and the Sciences

NELSON · Accelerated Testing, Statistical Models, Test Plans, and Data Analyses

NELSON · Applied Life Data Analysis

NEWMAN · Biostatistical Methods in Epidemiology

OCHI · Applied Probability and Stochastic Processes in Engineering and Physical Sciences

OKABE, BOOTS, SUGIHARA, and CHIU · Spatial Tesselations: Concepts and Applications of Voronoi Diagrams, *Second Edition*

OLIVER and SMITH · Influence Diagrams, Belief Nets and Decision Analysis

PANKRATZ · Forecasting with Dynamic Regression Models

PANKRATZ · Forecasting with Univariate Box-Jenkins Models: Concepts and Cases

*PARZEN · Modern Probability Theory and Its Applications

PEÑA, TIAO, and TSAY · A Course in Time Series Analysis

PIANTADOSI · Clinical Trials: A Methodologic Perspective

PORT · Theoretical Probability for Applications

POURAHMADI · Foundations of Time Series Analysis and Prediction Theory

PRESS · Bayesian Statistics: Principles, Models, and Applications

PRESS and TANUR · The Subjectivity of Scientists and the Bayesian Approach

PUKELSHEIM · Optimal Experimental Design

PURI, VILAPLANA, and WERTZ · New Perspectives in Theoretical and Applied Statistics

PUTERMAN · Markov Decision Processes: Discrete Stochastic Dynamic Programming

*RAO · Linear Statistical Inference and Its Applications, *Second Edition*

RENCHER · Linear Models in Statistics

RENCHER · Methods of Multivariate Analysis, *Second Edition*

RENCHER · Multivariate Statistical Inference with Applications

RIPLEY · Spatial Statistics

RIPLEY · Stochastic Simulation

ROBINSON · Practical Strategies for Experimenting

ROHATGI and SALEH · An Introduction to Probability and Statistics, *Second Edition*

ROLSKI, SCHMIDLI, SCHMIDT, and TEUGELS · Stochastic Processes for Insurance and Finance

ROSS · Introduction to Probability and Statistics for Engineers and Scientists

ROUSSEEUW and LEROY · Robust Regression and Outlier Detection

RUBIN · Multiple Imputation for Nonresponse in Surveys

RUBINSTEIN · Simulation and the Monte Carlo Method

RUBINSTEIN and MELAMED · Modern Simulation and Modeling

RYAN · Modern Regression Methods

RYAN · Statistical Methods for Quality Improvement, *Second Edition*

SALTELLI, CHAN, and SCOTT (editors) · Sensitivity Analysis

*SCHEFFE · The Analysis of Variance

SCHIMEK · Smoothing and Regression: Approaches, Computation, and Application

SCHOTT · Matrix Analysis for Statistics

SCHUSS · Theory and Applications of Stochastic Differential Equations

SCOTT · Multivariate Density Estimation: Theory, Practice, and Visualization

*SEARLE · Linear Models

SEARLE · Linear Models for Unbalanced Data

SEARLE · Matrix Algebra Useful for Statistics

SEARLE, CASELLA, and McCULLOCH · Variance Components

SEARLE and WILLETT · Matrix Algebra for Applied Economics

SEBER · Linear Regression Analysis

SEBER · Multivariate Observations

SEBER and WILD · Nonlinear Regression

SENNOTT · Stochastic Dynamic Programming and the Control of Queueing Systems

*Now available in a lower priced paperback edition in the Wiley Classics Library.

*SERFLING · Approximation Theorems of Mathematical Statistics

SHAFER and VOVK · Probability and Finance: It's Only a Game!

SMALL and McLEISH · Hilbert Space Methods in Probability and Statistical Inference

STAPLETON · Linear Statistical Models

STAUDTE and SHEATHER · Robust Estimation and Testing

STOYAN, KENDALL, and MECKE · Stochastic Geometry and Its Applications, *Second Edition*

STOYAN and STOYAN · Fractals, Random Shapes and Point Fields: Methods of Geometrical Statistics

STYAN · The Collected Papers of T. W. Anderson: 1943–1985

SUTTON, ABRAMS, JONES, SHELDON, and SONG · Methods for Meta-Analysis in Medical Research

TANAKA · Time Series Analysis: Nonstationary and Noninvertible Distribution Theory

THOMPSON · Empirical Model Building

THOMPSON · Sampling, *Second Edition*

THOMPSON · Simulation: A Modeler's Approach

THOMPSON and SEBER · Adaptive Sampling

TIAO, BISGAARD, HILL, PEÑA, and STIGLER (editors) · Box on Quality and Discovery: with Design, Control, and Robustness

TIERNEY · LISP-STAT: An Object-Oriented Environment for Statistical Computing and Dynamic Graphics

TSAY · Analysis of Financial Time Series

UPTON and FINGLETON · Spatial Data Analysis by Example, Volume II: Categorical and Directional Data

VAN BELLE · Statistical Rules of Thumb

VIDAKOVIC · Statistical Modeling by Wavelets

WEISBERG · Applied Linear Regression, *Second Edition*

WELSH · Aspects of Statistical Inference

WESTFALL and YOUNG · Resampling-Based Multiple Testing: Examples and Methods for p-Value Adjustment

WHITTAKER · Graphical Models in Applied Multivariate Statistics

WINKER · Optimization Heuristics in Economics: Applications of Threshold Accepting

WONNACOTT and WONNACOTT · Econometrics, *Second Edition*

WOODING · Planning Pharmaceutical Clinical Trials: Basic Statistical Principles

WOOLSON and CLARKE · Statistical Methods for the Analysis of Biomedical Data, *Second Edition*

WU and HAMADA · Experiments: Planning, Analysis, and Parameter Design Optimization

YANG · The Construction Theory of Denumerable Markov Processes

*ZELLNER · An Introduction to Bayesian Inference in Econometrics

ZHOU, OBUCHOWSKI, and McCLISH · Statistical Methods in Diagnostic Medicine

*Now available in a lower priced paperback edition in the Wiley Classics Library.